The Theory of
Everything,
Solved

A simplified explanation of the
nature of the universe

Lawrence J. Wippler

iUniverse, Inc.
New York   Bloomington

The Theory of Everything, Solved
A simplified explanation of the nature of the universe

Copyright © 2009 Lawrence J. Wippler

iUniverse books may be ordered through booksellers or by contacting:

iUniverse
1663 Liberty Drive
Bloomington, IN 47403
www.iuniverse.com
1-800-Authors (1-800-288-4677)

ISBN: 978-0-595-51107-5 (pbk)
ISBN: 978-1-4401-2047-3 (cloth)
ISBN: 978-0-595-61767-8 (ebk)

Printed in the United States of America

iUniverse rev. date: 3/2/2009

Contents

LIST OF FIGURES   VII

PREFACE   IX

INTRODUCTION   XI

1. MAGNETIC MONOPOLES   1

2. MAGNETISM   10

3. ELECTRICITY   15

4. BUILDING BLOCKS   21
    THE ATOM   21
    MATTER   24
    HEAT   28
    LIGHT   31

5. ENERGY   35

6. FORCES   45
    ETHER   45
    GRAVITY   48
    CENTRIFUGAL FORCE   52

7. SPACE   56
    BLACK HOLES   56
    EDDY CURRENTS   60
    EARTH AND THE MOON   62
    THE SEASONS   69
    AURORA BOREALIS   76
    HIGH-FREQUENCY (HF) RADIO WAVE PROPAGATION   79

8. SUPERCONDUCTORS   84

CONCLUSION   91

GLOSSARY   93

ABOUT THE AUTHOR   97

Figures

1 A.	Magnetic lines of force	2
1 B.	Magnetic monopoles	4
1 C.	Magnetic field lines.	6
1 D.	Magnetic monopoles' no repel/attraction area.	7
1 E.	Magnetic monopoles' perpetual motion	7
2.	Weak magnetic lines of force	.11
3 A.	Atoms' lines of force: electricity	.16
3 B.	Separating monopoles.	.19
4 A.	Atoms' magnetic lines of force	.23
4 B.	A prism	.33
5.	Elements' magnetic field lines.	.36
6 A.	Ether	.46
6 B.	Gravity	.50
6 C.	Centrifugal force.	.54
7 A.	A black hole	.58
7 B.	Large- and small-scale magnetic lines of force.	.61
7 C.	The moon's orbit.	.66
7 D.	Earth's orbit	.67
7 E.	Earth's seasons.	.72
7 F.	The moon's tilt	.74
7 G.	Radio wave propagation.	.81
8.	A superconductor	.87

Preface

The Theory of Everything, Solved was written to provide an alternate understanding of the inner workings of the atom. For many years, scientists have tried to unite the four fundamental forces—the strong and weak nuclear forces, gravity, and electromagnetism—without success. Many have tried uniting known theories, such as general relativity with quantum mechanics, string theory, and even the standard model. These theories differ in many ways, and it seems difficult, if not impossible, to find a single link to connect these theories together.

Many scientists believe that a single "theory of everything" must contain one or more of the present-day theories, or perhaps an equation. Albert Einstein spent the last years of his life searching without success for an equation that would unite the fundamental forces. With so many scientists working on many theories, perhaps they have overlooked the simple theories, the ones that do not require complex equations or have to unite one theory with another.

Having an interest in physics as a hobby, I began to formulate my own theory that would unite the fundamental forces. I first began by not making the assertion that the solution was an equation or had to unite with an existing theory. At this point, I set aside all known theories and equations. I felt a new theory had to be developed, but given this goal, where would one start? I began this quest for a new theory by working with magnets. They seemed to be very simple—opposites attract and likes repel—but what makes a magnet a dipole? No one seemed to know just how magnets actually worked, so I came up with my own theory.

My theory uses only three particles: north and south magnetic monopoles and a particle of matter that represents an element. The characteristics of these particles never change; the only thing that changes is how these monopoles move. With this basis, I then developed theories that would explain how these particles interact with each other and how they are able to create all forms of energy, including magnetism and gravity.

My theories explain how the fundamental forces of the universe are united by these particles and how they all interact with each other. This book does not answer all of the questions that scientists have asked about the universe; it only intends to show how this three-particle theory unites the fundamental forces of nature. Even so, my three-particle theory can answer all those questions, including the world's unsolved mysteries and paranormal phenomena. To help you understand this new theory, you must now set aside all presently known theories and laws of physics. Then look at the simplicity of the three-particle theory.

Introduction

I have always had an interest in physics as a hobby and have often wondered why the universal "theory of everything" has never been solved. A theory of everything (TOE) is a hypothetical one that, using theoretical physics, fully explains and links together all known physical phenomena. It is the Holy Grail of physics: one set of principles to unite the fundamental forces.

The standard model of particle physics is currently the best description of all experimental data. This theory explains the three major interactions of elementary particle physics—the strong interaction responsible for nuclear forces, the weak interaction responsible for radioactive decay, and the electromagnetic interaction. Nevertheless, there are reasons to believe that there are phenomena that are not accurately described by this theory. This model uses many elementary particles to explain things that occur in nature. But why so many particles? My personal belief is that all things should be kept simple. With this in mind, I began my quest to solve the theory of everything.

I began by keeping things simple. I would not use the standard model's laws of physics, because that model uses so many elementary particles. Second, string theory has its many dimensions. These are both complex theories, which meant I had to start from the beginning by creating my own theory. But where does one start when trying to solve the theory of everything?

After giving this some thought I decided to begin with a magnet. Magnets had always been fun to play with. I often asked why some things stuck to them and other things did not. This was always puzzling to me, and no one I asked seemed to know the answer. Then I asked myself, how would I create a magnet? I took a good look at the magnet and asked, what is it made of? What are those invisible magnetic lines of force that surround a magnet? Why does it have north and south magnetic poles? I theorized that there must be more to the magnet than what the eye could see.

My research concluded that magnets must have point-like particles to create their magnetic poles. In 1931, Paul Dirac provoked new interest in the possibility that monopoles exist by tying them to the phenomenon of electric charge quantization. However, he did not know how to find the monopoles or how they created a magnetic field. My hypothesis is that each north magnetic monopole orbits through a core in the opposite direction of the south magnetic monopole; this would explain why a magnet is a dipole.

Through my research, I have created theories on electricity, gravity, light, magnetism, and the strong and weak nuclear forces using only these north and south magnetic monopoles.

All forms of energy, including gravity, are composed of these magnetic monopoles.

Imagining a time before the standard model and general relativity, and before the laws of physics that we know of today, will help you understand these new theories of how all things in nature are united. My theories use only three particles: north and south monopoles, which have the ability to attract and repel each other, and particles of matter that represent an element. These are all that is needed to explain the four fundamental forces of nature.

1. Magnetic Monopoles

I will begin this exploration into the theory of everything by explaining, using present-day laws of physics, how magnets work. I will then explain how magnets work under my new theory, which uses magnetic monopoles, and how to isolate monopoles. People have known about electricity and magnetism for centuries. In time, scientists found that there are two types of electric charges: positive and negative. These opposite charges attract each other, and the electric charge is quantized. That is, all electric charges are multiples of an elementary electric charge found on the electron. As for magnetism, the ancient Greeks knew that certain minerals attracted iron and other pieces of the same mineral. About a thousand years ago, the Chinese noticed that a magnetized needle always points in the same direction and thus can be used for navigation.

Unlike electric charges, which can be isolated, magnetic materials always have two poles (called north and south after the directions they point to on Earth). If one breaks a magnet into two pieces, each smaller piece will again have both a north and a south pole. It is therefore apparently impossible to isolate a single magnetic pole—only the combination of north and south poles (called a dipole) seems to exist.

The absence of a single magnetic charge (called a monopole) makes the laws of electricity and magnetism different. This lack of symmetry has bothered physicists for decades. We now know of two distinct methods of generating a magnetic field. We can either use a permanent magnet, such as a bar magnet, or we can run an electric current through a coil of wire. Are these two methods fundamentally different, or are they somehow related to each other? Let us investigate further.

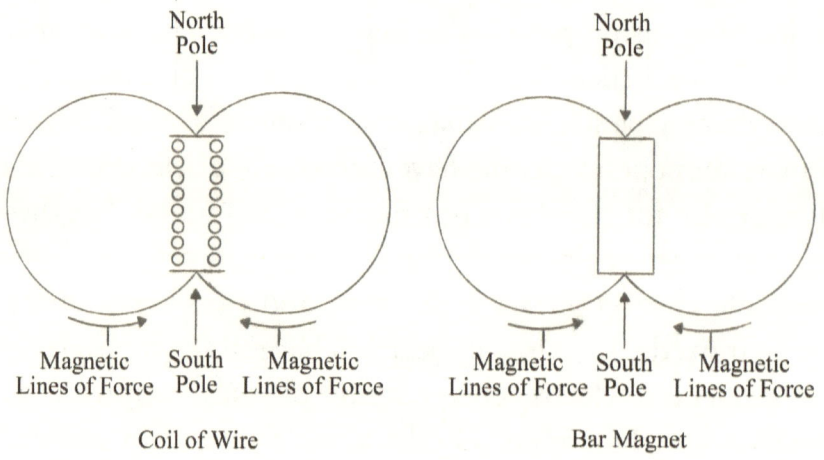

Fig 1 A.

Magnetic lines of force

As illustrated in figure 1 A, the external magnetic fields generated by a coil of wire and a conventional bar magnet are remarkably similar in appearance. Incidentally, these fields can easily be mapped out using iron filings. My first hypothesis is that the field of a bar magnet is produced by electric currents that flow around the outside of the bar magnet in a counterclockwise direction as we look along the magnet from its north to its south pole. There is no doubt, by analogy with a coil of wire, that such currents would generate a magnetic field.

My second hypothesis is that the field is produced by a positive magnetic monopole located close to the north pole of the magnet in combination with a negative monopole of equal magnitude located close to the south pole of the magnet. But what, exactly, is a magnetic monopole? Well, it is the magnetic equivalent of an electric charge. For example, a positive magnetic monopole is an isolated magnetic north pole.

We would expect magnetic field lines to radiate away from such an object, just as electric field lines radiate away from a positive electric charge. Likewise, a negative magnetic monopole is an isolated magnetic south pole. We would expect magnetic field lines to radiate toward such an object, just as electric field lines radiate toward a negative electric charge.

The magnetic field patterns generated by both types of monopole are sketched in figure 1 B.

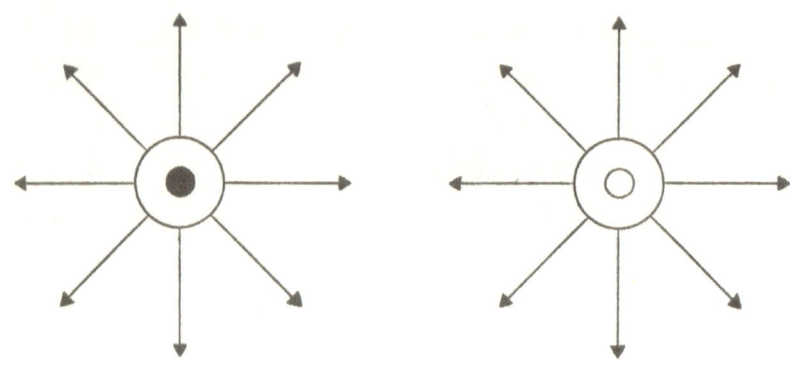

North Magnetic Monopole   South Magnetic Monopole

Fig 1 B.

Magnetic monopoles

We now have two hypotheses to explain the origin of the magnetic field of a bar magnet. What experiment could we perform in order to determine which of these two hypotheses is correct? Well, suppose that we break our bar magnet in two. What happens according to each hypothesis?

If we cut a coil of wire in two, then we just end up with two shorter coils of wire. So according to our first hypothesis, if we break a bar magnet in two, then we just end up with two smaller bar magnets. However, our second hypothesis predicts that if we break a bar magnet in two, then we end up with two equal and opposite magnetic monopoles. Needless to say, the former prediction is in accordance with experiment, whereas the latter most certainly is not.

Indeed, we can break a bar magnet into as many separate pieces as we like, and each piece will still act like a smaller, but otherwise equivalent, bar magnet. No matter how small we make the pieces, we cannot produce a magnetic monopole. In fact, nobody has ever observed a magnetic monopole experimentally, which leads most physicists to conclude that magnetic monopoles do not exist. Thus, we can conclude that the magnetic field of a bar magnet is produced by electric currents flowing over the surface of the magnets. However, what is the origin of these currents?

Perhaps there is one more hypothesis to consider. If you take a coil of wire and move a bar magnet past it, you get an electric current. This is how electricity is generated in a generator. So let us go one step further; if the movement of this bar magnet past this coil of wire produces electricity, then why would you have protons and electrons—which have an electric charge—coming out of this coil of wire? We all know that a bar magnet has a magnetic field, not an electric charge. Then why did the protons and electrons come out of the coil of wire? Perhaps they are not protons and electrons, but rather the north and south magnetic monopoles that Paul Dirac theorized existed in 1931 (see Introduction).

That is my hypothesis—that the proton and electron are really Dirac monopoles. The proton is actually a north magnetic monopole, and the electron is actually a south magnetic monopole.

In addition, just how would these magnetic monopoles behave? For example, the bar magnet is considered a dipole with its north magnetic pole at one end and south magnetic pole at the other. However, my hypothesis considers using two directions to create a magnetic field, as illustrated in figure 1 C.

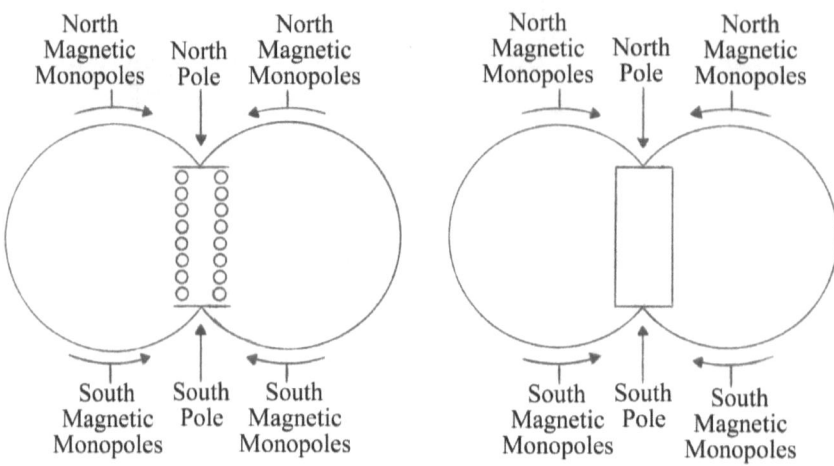

Fig 1 C.

Magnetic field lines

Figure 1 C shows that a magnet's north magnetic monopole would travel from the south pole of the bar magnet to the north pole of the bar magnet, and just the opposite would occur for the south magnetic monopole. Now both magnetic monopoles are moving in opposite directions. The north magnetic monopole will attract the south magnetic monopole in perpetual motion, creating a magnetic line of force. Without this perpetual motion of magnetic monopoles, you cannot have a magnetic field.

This perpetual motion is attained by the magnetic attraction between the north and south magnetic monopoles. It is my hypothesis that each monopole has an area that does not attract or repel. This area is close to the surface of the monopole (see figure 1 D). When north and south monopoles mutually attract, their magnetic attraction brings them closer together. But once these opposite monopoles get very close to each other, they no longer have any attraction for each other.

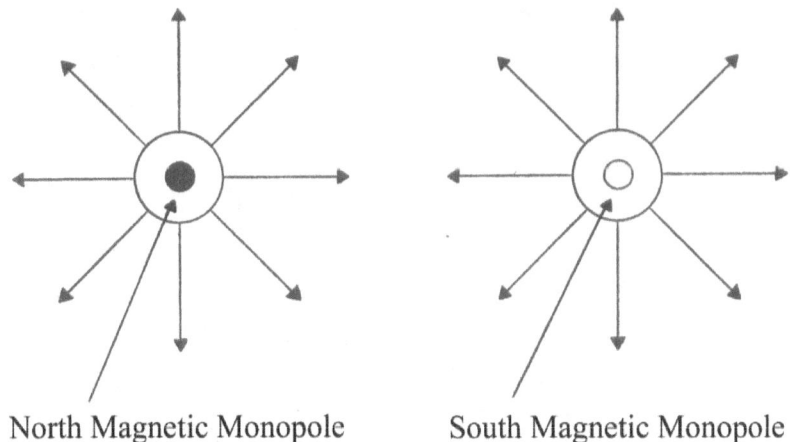

North Magnetic Monopole   South Magnetic Monopole

Fig 1 D.

Magnetic monopoles' no repel/attraction area

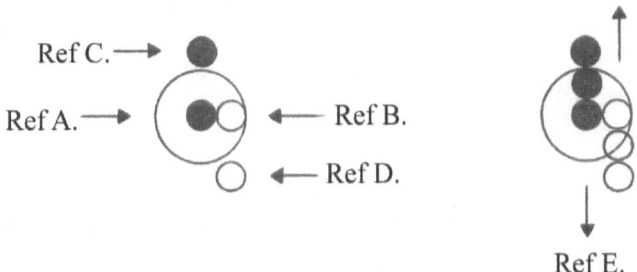

Fig 1 E.

Magnetic monopoles' perpetual motion

Figure 1 E, reference A shows the area of no repulsion or attraction. Any monopole in this area will experience neither an attractive nor repulsive force. Reference B shows a monopole in this area. That south monopole will then attract the north monopole at the area shown in reference C and bring it closer to the north monopole at reference A. Because reference A is a north monopole, it will repel the force of the north monopole already at reference A. The south monopole at reference B will then move closer to reference C, and the attractive force of reference A will keep the south monopole in the no repel/attraction area, bringing the north monopole at reference C into the no repel/attraction area of reference A. The attractive force of reference B will keep the north monopole near the north monopole of reference A, in the no repel/attraction area. Now the south monopole in reference D will take the place of the monopole that was in reference B, and this cycle will continue in perpetual motion, as shown in reference E.

It's this perpetual motion that creates a magnetic line of force. In bar magnets, this line of force also creates a north and south magnetic pole, making the bar magnet a dipole. figure 1 C shows the perpetual motion of the north and south magnetic monopoles that create the bar magnet's north and south magnetic poles. This is accomplished by the mutual attraction of the north and south magnetic monopoles. The south monopoles create a magnetic north pole by attracting all of the north monopoles in the same direction. Likewise, the north monopoles create a magnetic south pole by attracting all of the south monopoles in the same direction. This is why you continue to get a magnetic dipole when you break a bar magnet in half—since the perpetual motion of the north and south monopoles that make up the magnetic lines of force is unaffected by breaking the bar magnetic in half.

In conclusion, magnetic monopoles have already been discovered; they have been mislabeled as the proton and the electron. When you move a bar magnet past a coil of wire, it will attract and repel monopoles, which have a magnetic charge. Having an electric charge, protons and electrons will not be affected by the magnetic charge of a bar magnet and therefore cannot be coming out of the coil of wire.

It's these north and south magnetic monopoles that produce electricity, not protons and electrons. This explains why you get a magnetic field around a coil of wire when an electric current is passed through it. Electricity is created by separating the north and south magnetic monopoles into concentrated streams. This separation of magnetic monopoles is accomplished using a bar magnet, which has a magnetic charge and thus attracts the opposite monopoles in the same direction that the bar magnet is moving. When these concentrated streams of north and south magnetic monopoles recombine in a coil of wire, they will again create a magnetic field producing a dipole, because this is where they had originated before being separated by a bar magnet.

These are also the same north and south magnetic monopoles that create the magnetic lines of force in a bar magnet, which are responsible for producing all forms of energy in the electromagnetic spectrum, including gravity. Using only north and south magnetic monopoles and a particle of matter that represents an element, you can build an atom. With this three-particle atom, you can unite the fundamental forces of nature.

2. Magnetism

In physics, magnetism is one of the phenomena by which materials exert attractive or repulsive forces on other materials. Some well-known materials that exhibit easily detectable magnetic properties (called magnetic metals) are nickel, iron, cobalt, and their alloys. However, all materials are influenced to a greater or lesser degree by the presence of a magnetic field.

My hypothesis, as stated in the previous chapter, is that magnetism is created by north and south magnetic monopoles moving in opposite directions in an orbit through a common core. North and south magnetic monopoles continuously attract each other, creating perpetual motion. This is the same movement of magnetic monopoles that orbit through an atom's nucleus, creating a magnetic line of force and creating a magnetic dipole. A bar magnet has the same movement of magnetic monopoles as the atom, but on a much larger scale.

It is well known that magnets only attract certain metals and not others. Metals that are attracted to magnets are made from ferrous atoms, and metals that are not attracted to magnets are made from non-ferrous atoms. The difference between the two is that ferrous atoms have larger and weaker magnetic lines of force that orbit through their nuclei. Note that all ferrous and non-ferrous atoms are magnetic dipoles.

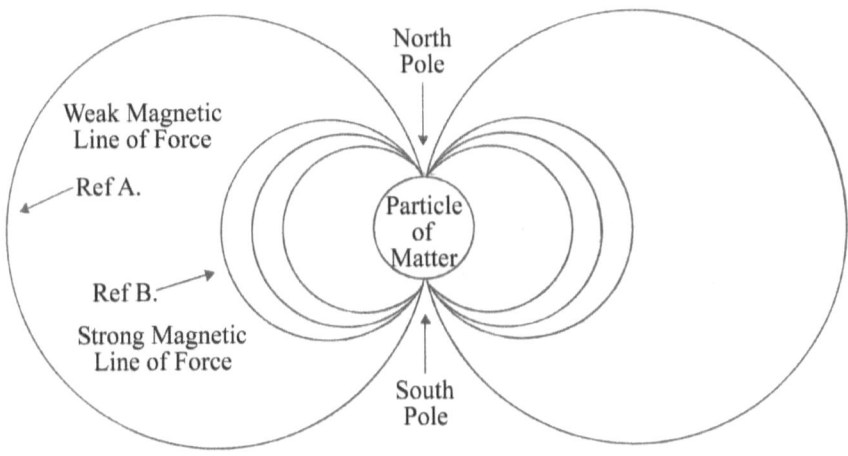

Fig 2.

Weak magnetic lines of force

Figure 2, reference A shows this outer magnetic line of force, which can be easily combined with other ferrous atoms to form a much larger magnetic line of force. This effect is known as ferromagnetism and is why you are able to create a bar magnet from any ferrous atom. Non-ferrous atoms do not have this larger and weaker magnetic line of force orbiting through their nuclei and therefore cannot be magnetized. Non-ferrous atoms have magnetic lines of force that are much stronger and therefore cannot be separated and combined to create this larger magnetic line of force that all bar magnets have. See reference B.

To magnetize metal made of ferrous atoms, all you have to do is combine the larger and weaker magnetic lines of force (which orbit through the atoms' nuclei) of all the atoms in the metal. This will create a much larger magnetic line of force. This is done by separating the north and south magnetic monopoles from the atoms' weaker magnetic lines of force and combining them with the magnetic lines of force from the other atoms in the metal. The stronger magnetic lines of force that orbit close to the atoms' nuclei are unaffected because they are much stronger than the larger, but weaker, magnetic line of force.

To separate the north and south magnetic monopoles from this weaker magnetic line of force, you must use a bar magnet in the same way that you would create electricity; just move the bar magnet's north pole in one direction over the ferrous metal. This will separate the north and south magnetic monopoles in this weak magnetic line of force. This is accomplished by the bar magnet's north pole, which attracts all of the south magnetic monopoles and concentrates them in one side of the metal, in turn causing the weak magnetic line of force of the atom to combine with the other weak magnetic lines of force from all of the other atoms in the ferrous metal. Finally, this creates a much larger magnetic line of force that will extend far beyond the ferrous metal. This large magnetic line of force creates a magnetic field, producing a magnetic dipole.

Note that if you were to break your newly created bar magnet in half, you would continue to have a magnetic dipole because (as stated earlier) the north and south magnetic monopoles that create its magnetic lines of force will be unaffected by the break.

The strength of a magnetic field is influenced by its temperature in that the field becomes stronger when it is cooled and weaker when it is heated. The same thing happens to the atom, which is also a magnetic dipole. The temperature at which a bar magnet loses its magnetic field is known as its Curie temperature. My hypothesis is that heat is the random movement of north and south magnetic monopoles. It's this movement of magnetic monopoles, this heat, that interferes with the perpetual motion of the magnetic lines of force that produce a magnetic dipole. The more interference that is caused by heat, the slower the perpetual motion and the weaker the magnetic field becomes.

Paramagnetism is a form of magnetism that occurs only in the presence of an externally applied magnetic field. Paramagnetic materials are attracted to magnetic fields and therefore have a relative magnetic permeability. This is caused by atoms' magnetic lines of force that are too strong to be separated and subsequently combined like the weak outer magnetic lines of force in ferrous atoms. However, they can be combined when in the presence of a strong external magnetic field. Only when these non-ferrous atoms become hot will their magnetic lines of force become weaker, just like the outer magnetic lines of force in ferrous atoms do. This allows the atoms in the metal to become magnetized only in the presence of a strong external magnetic field. When this magnetic field is removed, the atoms' magnetic lines of force will again return to their original orbits through the atoms' nuclei.

Antiferromagnetic materials are materials that exhibit ferromagnetism at a low temperature and become paramagnetic above a certain temperature; the transition temperature is called the Néel temperature. Above the Néel temperature, the material is typically paramagnetic. The Néel temperature is the point at which the atoms' magnetic lines of force become too strong for the external magnetic field to combine the weaker magnetic lines of force. This will prevent the material from becoming magnetized, and instead the material will exhibit paramagnetic properties.

Magnetism is created by the perpetual motion of magnetic monopoles. These monopoles are the cosmic force; they are what bind atoms together to create matter and also to produce a magnetic field, which creates a magnetic dipole. The strength of atoms' magnetic lines of force changes with their distance from the atom's core. The smaller the diameter of the magnetic lines of force, the closer they are to the atoms' core and the stronger the atoms' magnetic field becomes. Heat is also a factor that determines the strength of the atoms' magnetic line of force. An increase in temperature will decrease the strength of the atoms' magnetic lines of force, weakening the bond between atoms. This weakening is caused by the random movement of the north and south magnetic monopoles, which interfere and slow down the perpetual motion of the atoms' magnetic lines of force.

3. Electricity

Electricity is defined as the flow of electrical power or charge. In order to understand how an electric charge moves from one atom to another, you must understand the atom and how it works.

Everything in the universe is made from atoms—every star, every tree, every animal. The human body is made of atoms. Air and water are, too. Atoms are the building blocks of the universe. Atoms are so small that millions of them can fit on the head of a pin. As small as they are, atoms are made of even smaller particles. The center of an atom is called the nucleus; under my theory this nucleus is composed of a particle of matter representing an element from the periodic table. The atom is also made of particles called north magnetic monopoles and south magnetic monopoles. The north magnetic monopole carries a north magnetic charge, and the south magnetic monopole carries a south magnetic charge. Opposite charges attract each other, and like charges repel each other. These magnetic monopoles are very small and are what create atoms' magnetic lines of force, producing a magnetic dipole.

To summarize, atoms are small magnetic dipoles composed of only three particles. The north and south magnetic monopoles that orbit through an atom's core are the same monopoles that create electricity when they have been separated into concentrated streams.

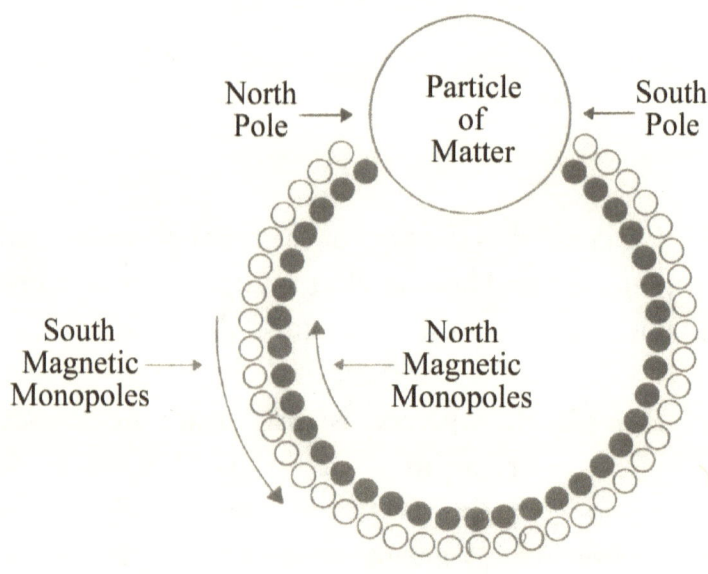

Fig 3 A.

Atoms' lines of force: electricity

Figure 3 A shows what an atom would like if it were large enough for you to see it. In the center of the atom is the nucleus. In this nucleus is a particle of matter that represents an element, which can be one of the many elements from the periodic table. The north and south magnetic monopoles are particles that orbit through the atom's nucleus in opposite directions. These magnetic monopoles are in perpetual motion, creating a magnetic line of force and producing a magnetic dipole. For each element, these magnetic lines of force are in different positions relative to their atom's nucleus. Only one magnetic line of force is shown; however, an atom has many.

Electricity is produced by separating the north and south magnetic monopoles from their atom's magnetic lines of force into concentrated streams. The concentrated stream of north magnetic monopoles is commonly known as positive electricity, and these monopoles move in a clockwise direction. This concentrated stream of north monopoles moves in the same direction as the north monopoles that create one half of the atom's magnetic line of force. This is because the stream of north magnetic monopoles originated from the atom's magnetic lines of force, which were then replaced by the many surplus magnetic monopoles from the ether. The ether, which occupies all space, is composed of both north and south magnetic monopoles and is the medium through which all energy propagates.

Likewise, the concentrated stream of south magnetic monopoles is commonly known as negative electricity, and these monopoles move in a counterclockwise direction. This is the same direction as the south monopoles in the atom's magnetic line of force, because this is where they had originated. This magnetic line of force is what creates the atom's magnetic field, which binds atoms together to create matter.

To generate electricity, all you have to do is separate the north and south magnetic monopoles from the atoms' magnetic lines of force to create concentrated streams. This is accomplished by moving a bar magnet past a coil of wire.

When concentrated streams of north and south magnetic monopoles run against each other in a wire, also known as electricity, the north and south magnetic monopoles will recombine. Because they are moving in opposite directions, they will also create a magnetic line of force. This is the same magnetic line of force from which the monopoles had originated.

Now let's look at how electricity is created when using a generator. A generator is a device that converts mechanical energy into electrical energy. The process is based on the relationship between magnetism and electricity. In 1831, English physicist Michael Faraday discovered that when a magnet moved inside a coil of wire, magnetic current flows in the wire. A typical generator at a power plant uses an electromagnet—a magnet produced by electricity—not a traditional magnet. The only difference between the two types of magnets is how an electromagnet is created. Electromagnets are created by supplying separated north and south magnetic monopoles (electricity) into a coil of wire. This recombines the separated monopoles, allowing them to once again create their original magnetic lines of force.

A generator has a series of insulated coils of wire that form a stationary cylinder. This cylinder surrounds a rotary electromagnetic shaft. When the electromagnetic shaft rotates, it induces a small magnetic current in each section of the wire coil. Each section of the wire becomes a small, separate electric conductor. The small currents of individual sections are added together to form one large current. This current is the electric power that is transmitted from the power company to the consumer.

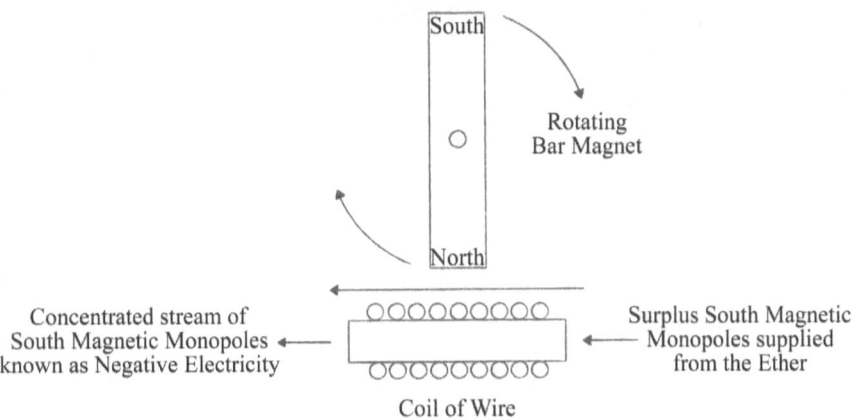

Fig 3 B.

Separating monopoles

Figure 3 B shows how electricity is generated in a coil of wire. This is accomplished by the atoms that make up the wire; each atom is a small magnetic dipole with its own north and south magnetic monopoles orbiting through its nucleus. When the north pole of a bar magnet passes by the magnetic field of the atom, it attracts all of the south magnetic monopoles that are in orbit through the atom's nucleus. When a south magnetic monopole is pulled away from the atom's magnetic lines of force by the north pole of the bar magnet, it is then replaced by one of the many surplus south magnetic monopoles from the ether. This process concentrates the south magnetic monopoles in one end of the coil of wire. If these south magnetic monopoles are not used as electricity right away, they will go back into the ether as surplus monopoles. After the north pole of the bar magnet has passed, the south pole of the bar magnet then concentrates the north magnetic monopoles in the coil of wire the same way the south magnetic monopoles were concentrated. This is how alternating currents are created in a generator. When these separated magnetic monopoles are again recombined in a wire, they create a magnetic line of force, producing a magnetic dipole.

Because electricity originates from atoms' magnetic lines of force, it also uses these lines of force to propagate through. This is because the north and south magnetic monopoles that create electricity are moving in the same direction as the atoms' magnetic lines of force. For this reason, electricity is able to propagate through metal better than air, since the atoms in metal are much closer together and have a much stronger magnetic field that binds them together. The stronger and closer the atoms' magnetic field is, the less resistance a material of that atom has to electricity. To reiterate, in a gas such as air, the atoms have a much weaker magnetic field and are farther apart. Since electricity uses the atoms' magnetic field to propagate, air has a much higher resistance to electricity than metal.

Electricity is nothing more than north and south monopoles that have been separated. These are the same monopoles that create a magnetic dipole and all forms of energy. The separation of these monopoles is accomplished using a magnetic dipole. After these monopoles are recombined in a coil of wire, they will again create a magnetic field.

4. Building Blocks

The Atom

I have concluded that the atom consists of only three parts: the nucleus and north and south magnetic monopoles. This model of the atom differs greatly from the standard model, which has many elementary particles.

In the nucleus of the atom, there is a particle of matter that represents an element; this can be any element from the periodic table. It's this element that gives the atom its own unique properties in that each unique element is able to hold a different amount of north and south magnetic monopoles.

These magnetic monopoles are what orbit through the atom's nucleus, creating its magnetic lines of force. The magnetic lines of force are all located in different positions relative to the atom's nucleus. You may think of this as the atom's DNA, with no two elements having the same DNA.

Because each element is able to hold a different amount of magnetic lines of force, each element also creates a stronger or weaker magnetic field than other elements. Elements that have a large number of magnetic lines of force also have a high atomic number; this number is the amount of magnetic lines of force that an atom is able to hold.

Gravity is composed of north and south magnetic monopoles that are moving alongside each other and in the same direction relative to space. Therefore, gravity attracts the north and south magnetic monopoles that orbit through the atom's nucleus, causing atoms with a large atomic number or amount of magnetic lines of force to become very heavy (called heavy elements). Elements with a low atomic number do not weigh as much because they do not have as many magnetic lines of force orbiting through their atoms' nuclei. Therefore, in lighter elements, there are fewer magnetic lines of force for gravity to attract.

The atom also transforms energy by changing the direction of these magnetic monopoles using the atom's magnetic lines of force. The greater the amount of magnetic lines of force the atom has, the more energy the atom is able to transform. If you have an element with a large atomic number, that element is able to transform more energy, and thus has a greater mass than an element with a low atomic number.

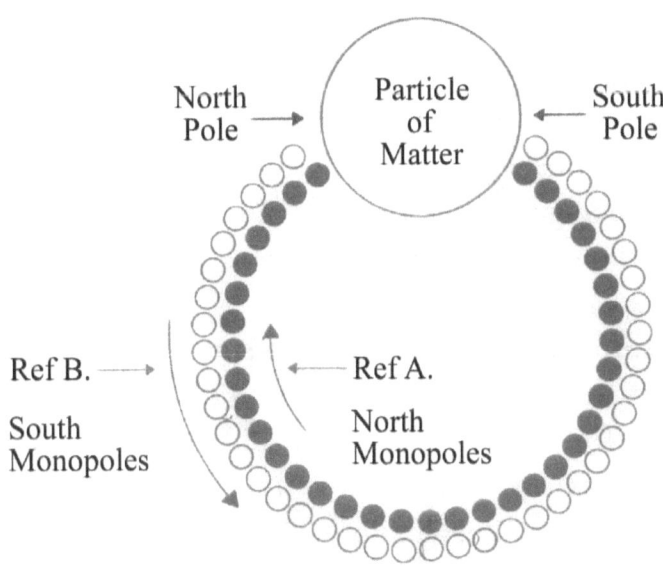

Fig 4 A.

Atoms' magnetic lines of force

Figure 4 A, reference A shows how north magnetic monopoles move in a clockwise direction through the atom's nucleus. These magnetic monopoles are so small they can pass through everything. North monopoles make up one half of the atom's magnetic line of force, which creates the atom's magnetic field and produces a magnetic dipole. This dipole is created by the movement of these monopoles. The north magnetic monopoles that are moving in a clockwise direction attract all excess south monopoles to move in the opposite direction, creating a magnetic south pole. The south magnetic monopoles create a north magnetic pole by attracting all excess north monopoles in the opposite direction. The end result is a magnetic dipole with both north and south magnetic poles.

Reference B shows how the south magnetic monopoles move in a counterclockwise direction through the atom's nucleus. This is because south magnetic monopoles always attract north magnetic monopoles in perpetual motion. This movement of north and south magnetic monopoles creates a magnetic line of force, giving the atom its magnetic field and producing a magnetic dipole. Only one magnetic line of force is shown in this diagram; however, atoms have many magnetic lines of force.

To summarize, atoms are just small magnetic dipoles. The particle of matter in the atom's nucleus represents an element. This particle determines the number of magnetic lines of force and their relative positions from the atom's nucleus. Further, the strength of an atom's magnetic field changes with temperature; the hotter it gets, the weaker its magnetic field becomes. When the atom is cooled to extreme low temperatures, its magnetic field becomes very strong.

Matter

Matter is the binding of two or more atoms together. Depending on the strengths of the atoms' magnetic fields, the matter forms a solid, liquid, or gas. Magnetic monopoles are the binding force that holds these atoms together. During the process of covalent bonding, monopoles are often released as heat; however, during the process of ionic bonding, they are absorbed, creating a cooling effect.

The bonding of two atoms together is accomplished by the north and south magnetic monopoles, which are in perpetual motion, creating a magnetic line of force and producing a magnetic dipole.

Every element is different in that it is able to have a different number of magnetic lines of force orbiting through its nucleus. These lines of force are circular, and the magnetic lines of force that are closest to the element's core are the smallest diameter and have the strongest magnetic field. The farther you get from the element's core, the weaker this magnetic field becomes and the larger its diameter will be.

Different elements have different amounts of magnetic lines of force that orbit through their atoms' nuclei. These lines of force are not all in the same positions relative to the element's nucleus and therefore are not all the same diameter. When two atoms of different elements combine, they will often share one or more of their magnetic lines of force. These lines of force can only combine if they are the same diameter and the same distance from the elements' cores.

When these magnetic lines of force combine, the area that is shared gives off the north and south magnetic monopoles that are no longer needed. This movement of monopoles will be random, because the lost monopoles are not replaced by new monopoles. This random movement of monopoles creates what is known as heat, which interferes with the perpetual motion of the atoms' magnetic lines of force, weakening their magnetic field. The more lines of force that combine between two elements, the stronger the bond. This process of bonding magnetic lines of force to create matter is known as covalent bonding.

However, a bond between two elements can also be created by the atoms' magnetic lines of force that are not exactly the same diameter or the same distance from the elements' cores, as long as the difference is small. In this case, as long as the small difference in distance allows room for a line of force between the two, they, too, can combine together to form matter. This is known as ionic bonding.

When these two elements combine, they do not release any excess north and south monopoles as heat because the magnetic lines of force do not share any monopoles. This is due to the very slight difference in diameter of their magnetic lines of force, creating an ionic bond. For these two elements to fully combine, they must acquire more north and south monopoles. The source of these magnetic monopoles is the pool of excess monopoles that are moving at random, known as heat. Therefore, when these excess monopoles are drawn into the two elements that are combining, the result is a lower temperature. Remember that heat is excess north and south magnetic monopoles that interfere with atoms' magnetic lines of force, causing them to slow down and weakening their magnetic field. Cold is just the opposite.

When excess monopoles are drawn into a magnetic line of force, this decreases the temperature of the elements that are bonding together by removing excess magnetic monopoles (heat). This also increases the speed of the monopoles' perpetual motion and creates a stronger magnetic field. After the two elements have completed their bonding process, these elements will again return to room temperature because they are no longer acquiring excess monopoles (heat).

There are three different states of matter—solid, liquid, and gas. The atoms in a solid have a strong magnetic field, binding the atoms together. The stronger the magnetic bond, the harder the solid matter is.

The atoms in a liquid are not bound together as tightly as the atoms in a solid are. This is due to the liquid's weaker magnetic field. This weaker magnetic field allows the atoms in a liquid to move relative to other atoms, and because of this, a liquid is able to flow and share some of the properties of a solid, such as an inability to be compressed. This property that both solids and liquids share is due to the distance between the atoms. In a solid, the atoms' magnetic field is very strong, binding the atoms closely together where they are not able to move relative to other atoms. In a liquid, the atoms' magnetic field is not quite as strong as a solid; however, the liquid's magnetic field keeps the atoms bound close together like a solid. The atoms that form a liquid are able to move relative to each other, creating a form of matter that cannot be compressed due to the close distance between the atoms, just like a solid.

Imagine that you have several rows of pool balls on a table, and these balls each have a magnetic field. If this magnetic field becomes very strong, the pool balls become bound together with a strong magnetic force that will not allow the pool balls to become separated or move around on the table. This configuration of pool balls would represent a solid. Now imagine the same table of pool balls, however with a weaker magnetic field binding them together. This weaker magnetic field allows the pool balls to move relative to each other, creating a form of matter called a liquid that cannot be compressed, just like a solid.

The atoms in a gas have a much weaker magnetic field than solids or liquids and are not tightly bound together. The distance between the atoms in a gas is very large because of this. This allows the atoms to move relative to each other and allows the atoms in a gas to be compressed. When a gas is compressed, it gives up some of its magnetic monopoles as heat. When this gas is allowed to expand, it reacquires the same magnetic monopoles it had released as heat, creating a cooling effect. This explains how a gas can change its state from a gas to a liquid and to a solid just by the changing the strength of its magnetic field—by adding or removing heat.

Heat

Heat is defined as the random movement of north and south magnetic monopoles. This random movement of monopoles slows down the perpetual motion of atoms' magnetic lines of force, weakening their magnetic field. Heat is the result of the random movement of excess north and south magnetic monopoles that have been released by a nuclear or chemical reaction.

Heat is generated when atoms of two different elements combine. These elements will share some of their monopoles, and the bonding process will release some of the monopoles because they are no longer needed. It's these excess magnetic monopoles that produce what is known as heat. The more magnetic monopoles that are released when these different elements combine, the more heat is produced.

These magnetic monopoles that create heat are moving past the magnetic monopoles that create the atom's magnetic lines of force. Because these magnetic monopoles (heat) are moving at random, as they enter the magnetic lines of force that orbit through the atom's nucleus, they attract the opposite monopoles that are moving toward the magnetic monopole that created the heat. Since these magnetic lines of force are circular, when the magnetic monopole of heat moves away from the atom's magnetic line of force, it will again attract the opposite magnetic monopoles (which created the atom's magnetic line of force). Now the north magnetic monopole of heat is moving in the opposite direction and it will become separated from the south magnetic monopole that is in orbit through the atom's nucleus. It's this attraction of the north and south magnetic monopoles that creates resistance when they are separated. This resistance of separation between the north and south magnetic monopoles slows down the perpetual motion of the atom's magnetic lines of force, weakening its magnetic field. Again, it's this magnetic field that binds atoms together to create matter.

Pressure can also generate heat. For instance, a gas is composed of atoms that have a weak magnetic field. This magnetic field cannot bind the atoms together as strongly as the atoms that create solid matter. Atoms that have a weak magnetic field also have a weak bond between them. This weak bond allows for increased distance between the atoms, creating a gas. Remember, the stronger the atoms' magnetic field is, the tighter the bond between the atoms.

When a gas is compressed, its magnetic field is also compressed. Since the atoms in a gas can only hold a limited amount of north and south magnetic monopoles, when these atoms are compressed, they have to give up some of their north and south magnetic monopoles as heat. The reason for this is that the atoms' outermost magnetic lines of force are no longer needed, so their magnetic monopoles that had created these magnetic lines of force are given off and begin to move at random, creating heat. These excess, "heat" monopoles also use the atoms' magnetic field to propagate. These magnetic monopoles that create heat move at random, unlike other forms of energy, which are circular, polarized with different wavelengths.

When heat propagates through a solid, it uses the solid's atoms' magnetic field, because heat is composed of the same stuff—north and south magnetic monopoles—that the atoms' magnetic field is made of. Since solids, such as metal, have atoms that are very close together and are bound by a strong magnetic field, heat is able to propagate with very little resistance. This is also the same way that electricity propagates; if a material will conduct electricity, it will also conduct heat.

When heat propagates through a liquid, it also uses the atoms' magnetic field. However, a liquid's magnetic field is weaker than a solid's because its atoms are not bound together as strongly. As a result, heat cannot propagate as well through liquids. But in a liquid, the atoms are able to move by convection, which allows heat to propagate at a much faster rate than metal.

In a gas, heat again uses the atoms' magnetic field to propagate. However, in a gas, the atoms' magnetic field is much weaker than a liquid, so the only way heat can propagate is through convection. If the atoms in a gas are unable to move, heat cannot propagate very well, so a gas that is not moving by convection will not be able to transfer heat. But if the gas is moving, heat will be able to propagate at a much faster rate.

Cold is just the opposite of heat. When an atom is cooled to very low temperatures, its magnetic field becomes very strong because heat interferes with the perpetual motion of the atoms' magnetic field. This is why things melt when they get hot—their magnetic field has become very weak and is unable to bind the atoms together.

When atoms are cooled to very low temperatures, their north and south magnetic monopoles no longer have any interference caused by heat to slow them down. This is why matter turns into a solid when their atoms become cold; their magnetic field has become very strong, and it's this magnetic field that binds the atoms together to create matter.

Heat always propagates from areas of hot to cold. Since heat causes atoms' magnetic field to become very weak, the weakened magnetic field does not attract the north and south magnetic monopoles that create all forms of energy, including heat. However, cold temperatures cause atoms' magnetic field to become very strong, which in turn causes a greater attraction for north and south magnetic monopoles. It's this difference in strengths of hot atoms' versus cold atoms' magnetic fields that causes heat to always flow from hot to cold.

Light

Light is an electromagnetic wave that you are able to see. This wave is created the same way that all other electromagnetic waves are created. Light waves are composed of north and south magnetic monopoles that are moving in opposite directions, just like the atoms' magnetic lines of force that created them. However, these monopoles are not in perpetual motion.

Light, like all other electromagnetic waves, is circular polarized. If you were able to look at a wave of light that was coming toward you, it would look just like the atoms' magnetic line force. A change in diameter would also mean a change in color. Light is composed of two particles: north and south monopoles. This gives the electromagnetic wave its particle-wave duality in that it acts as both a wave and a particle. Since light is an electromagnetic wave and composed of magnetic monopoles, it can be affected by gravity and magnetism. Light is thus able to be separated into its different wavelengths just by using magnetism.

A prism is a triangular piece of cut glass used to break white light up into its constituent colors. Visible light is composed of a spectrum of different colors, each with a distinct wavelength.

Prisms are made of atoms. These atoms are small magnetic dipoles in that they have both a north magnetic pole and a south magnetic pole. It's these poles that separate white light into its different wavelengths. Because light is circular polarized it takes the shape of a helix; the north magnetic monopole moves clockwise and the south magnetic monopole moves in the opposite direction—counterclockwise. Each stream of monopoles forms a separate helix. The north magnetic monopoles and the helix that they create complement the south magnetic monopoles in that they hold their opposite monopole in the no repel/attraction area that is near the surface of the monopole, just like the magnetic lines of force that transformed the stream of monopoles into the light wave. If one of these helices was removed, the other helix would become unstable because it would no longer have its opposite monopole to hold it in the no repel/attraction area. This would cause the remaining helix to slightly decrease in frequency and dissipate after a short distance. This effect is known as the Lamb shift and was discovered in April of 1947 by Willis Lamb Jr. while he worked with the Columbia radiation laboratory.

A prism separates colors by bending them by different degrees, creating a rainbow effect. Red, the lowest frequency visible to human eye, and bends less than violet, the highest frequency visible to the human eye. This means that red bends at a less severe angle than violet. All colors in between bend at different degrees, so the prism makes a rainbow out of white light. The bending of white light into its constituent colors is all accomplished by magnetism.

The prism takes the shape of a triangle. When light passes through a prism, the light wave encounters more atoms at the bottom half of the prism than at the top half due to the triangular shape of a prism, and this difference results in a magnetic imbalance. Because light is an electromagnetic wave that is composed of north and south monopoles, these monopoles become attracted to the atoms that make up the prism because atoms are also small magnetic dipoles. This creates a stronger magnetic attraction at the bottom half of the light wave than at its top half, causing the light wave to bend toward the thicker part of the prism. This is due to the increase in magnetic attraction created by the prism's atoms' magnetic field.

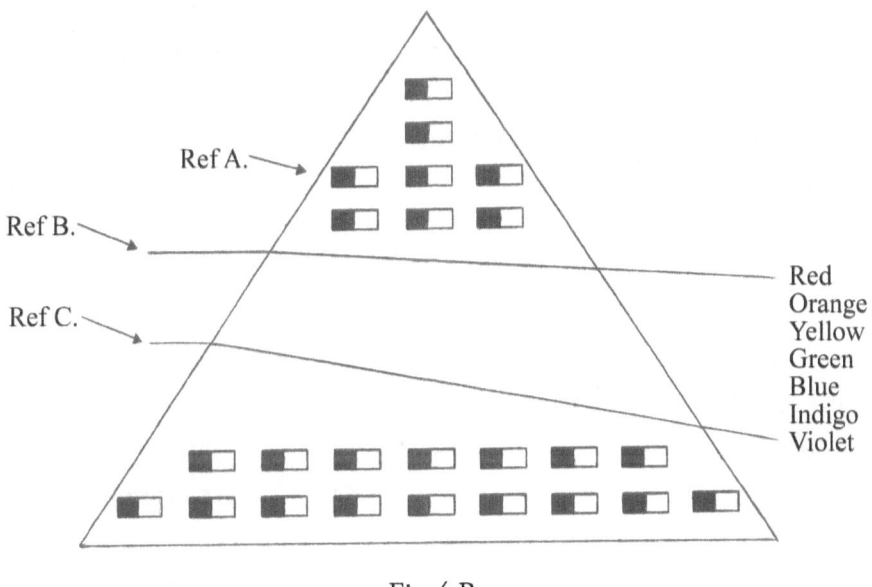

Fig 4 B.

A prism

In figure 4 B, reference A, the atoms that create the prism are shown as small magnetic dipoles. Because light is an electromagnetic wave that is composed of north and south magnetic monopoles, it will become attracted to the atoms' magnetic field. Since the light wave encounters more atoms at the bottom half of the wave than at its top half, a magnetic imbalance occurs. This causes the electromagnetic wave of light to become more attracted to the atoms at the bottom of the prism that those at the top, causing the light wave to bend at an angle. The angle at which the light wave is bent is dependent on its frequency.

Reference B shows that when a wave of light enters a prism that is oscillating at a low frequency, like the color red, it will only bend at a slight angle. Imagine a wave of light that is passing through a prism at a low frequency. As this electromagnetic wave oscillates, the bottom half of the wave encounters more atoms than the top half of the wave. This ratio of atoms encountered from the top to the bottom of the wave will change when the frequency of the electromagnetic waves changes. When a low-frequency electromagnetic wave of light (for example, the color red) exits the prism, the top half of the wave is no longer encountering the atoms that make up the prism. However, the bottom half is encountering the atoms that make up the bottom half of the prism. This bends the wave because of the magnetic attraction between the atoms that make up the prism and the north and south magnetic monopoles that compose the electromagnetic wave.

For reference C, now imagine a high-frequency electromagnetic wave of light (for example, the color violet). When this light exits the prism, the top half of the wave is no longer encountering atoms that make up the prism. However, the bottom half of the wave is still encountering the atoms the make up the prism. Because of its high frequency, the ratio of atoms from the top of the wave to the bottom the wave changes. This change causes a greater magnetic attraction at the bottom half of the wave than at the top half, causing the light wave to bend at a much greater angle than a low-frequency wave of light (like the color red) would. This is how light is separated into its constituent colors, creating a rainbow of color.

5. Energy

We use energy every day of our lives. In this chapter I will explain how energy is created and transformed from one form to another. Energy is all around us, and it takes many forms: light, gravity, heat, electricity—all are forms of energy.

Energy can only be transformed by the atom, though this can be done in several ways. Chemical and nuclear reactions, heat, and even electricity are able to transform one form of energy to another using the magnetic lines of force that surround every atom's nucleus.

My hypothesis defines energy as the movement of north and south magnetic monopoles. This movement of monopoles may be at random, creating heat, or perpetual motion, creating a magnetic dipole, or motion relative to space, creating gravity.

In the core of every atom is a particle of matter, which represents one of the many elements from the periodic table.

Each element has its own unique properties that differ from other elements, and no two elements are identical. The unique properties of each element are determined by the amount of magnetic lines of force that element is able to hold and the lines' relative positions from the core of the element. These are what create the different frequencies of the electromagnetic spectrum, as well. This difference also affects how the elements react with each other and how atoms transform energy from one form to another.

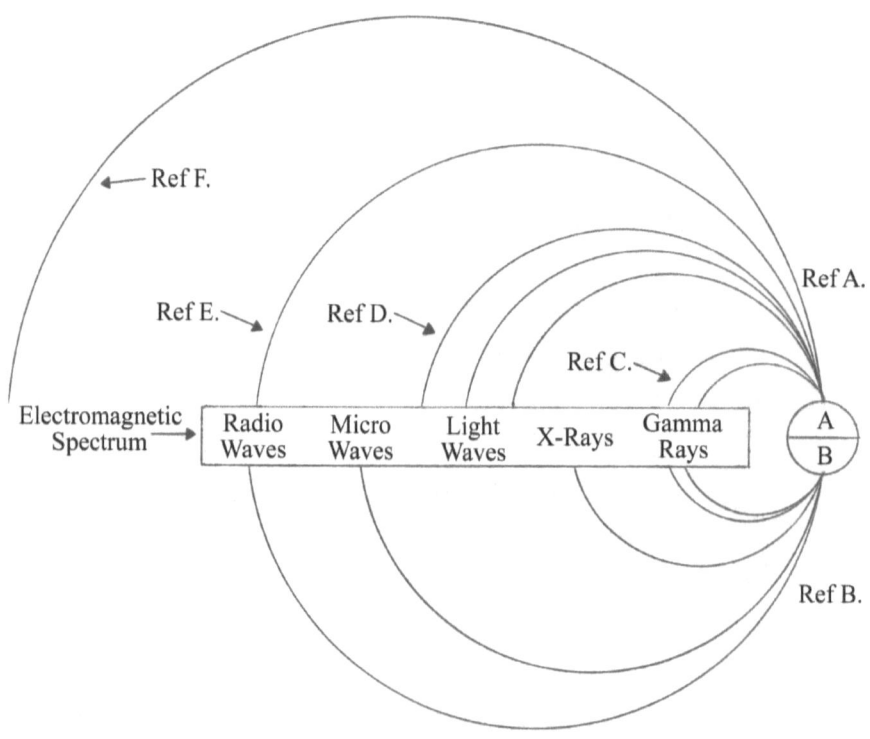

Fig 5.

Elements' magnetic field lines

Reference A of figure 5 represents element A and shows what an atom would look like if you were able to see it. Only a small portion of the element is shown in this illustration.

Element A has many magnetic lines of force, each of which represents a different frequency of the electromagnetic spectrum and also creates the atom's magnetic field, resulting in a magnetic dipole.

Reference B represents element B. Notice the difference between it and element A in terms of the number of magnetic lines of force that each element has and their relative positions from their atoms' nuclei. This is why atoms are able to transform energy into different frequencies of the electromagnetic spectrum; it's due to the relative positions and the diameters of their magnetic lines of force.

Reference C shows that the closer the magnetic lines of force are to the atom's nucleus, the smaller their diameter will be. A small-diameter magnetic line of force creates a short wavelength, high frequency, circular polarized electromagnetic wave in the gamma ray region of the electromagnetic spectrum. The farther the magnetic lines of force are from the atom's nucleus, the larger their diameter is. These large diameters create circular polarized, low-frequency waves in the microwave region of the electromagnetic spectrum.

All forms of energy are created this way, except magnetism and gravity, which are not electromagnetic waves. All electromagnetic waves created by the atom are circular polarized, composed of both north and south magnetic monopoles, and take the form of helices. These monopoles that create electromagnetic waves move in opposite direction of each other; however, they are not in perpetual motion. When used in an electromagnetic wave, they form a helix with the north monopole moving opposite of the south monopole.

Heat is the cause of the random movement of north and south magnetic monopoles; it's this random movement that slows down the perpetual motion of the atom's magnetic lines of force, weakening its magnetic field. The weakened field allows for an easier transformation of energy. However, when the atom is at extremely low temperatures and its magnetic lines of force become much stronger as discussed in previous chapters, to transform the magnetic line of force into an electromagnetic wave would require a stream of north and south magnetic monopoles moving at a much faster speed than the atom's magnetic lines of force.

When a stream of north monopoles (also known as electricity, see chapter 3) strikes an atom's magnetic lines of force, it always seeks its opposite, the south monopole. However, these south monopoles are moving in the opposite direction as the north monopoles, cancelling out the attraction between them. So the stream of north monopoles, or positive electricity, will strike the north monopoles in the atom's magnetic lines of force. The speed of this stream must exceed the speed of the north monopoles in the atom's magnetic lines of force to be able to transform energy. When this stream of north monopoles collides with the north monopoles in the atom's magnetic lines of force, the monopoles take the form of a helix having the same diameter and moving in the same direction as the line of force that produced them. This is how the atom transforms one form of energy to another.

Imagine a pool table with all the pool balls lined up in rows. When another ball strikes this row of pool balls, it will then transform its energy into the rows of pool balls, and a pool ball will emerge at the other end traveling in the same direction as the original ball that entered. Magnetic monopoles do exactly the same thing, but on a much smaller scale. Since an atom's magnetic lines of force are created from both north and south magnetic monopoles that are in perpetual motion, in order to have a complete transformation of energy, you must also use streams of south magnetic monopoles, also known as electricity.

This is an example of how energy is created in the electromagnetic spectrum. Light-emitting diodes (LEDs) are an efficient way of producing light. LEDs are semiconductors that use different elements to produce different colors of light. The colors of light vary in wavelength, just like the magnetic lines of force that produced them. Therefore, a red LED requires less voltage and current than a blue LED. This is a result of atoms' magnetic lines of force that produced the red color having a weaker magnetic field than the atoms producing a blue color. Having a weaker magnetic field means the red LED's atoms require less energy to transform the stream of monopoles into a circular, polarized electromagnetic wave than the blue LED's atoms require.

To complete this transformation of energy by an atom's magnetic lines of force, you must input an equal amount of energy into the atom's magnetic lines of force. To put it in easy terms, you cannot create a higher energy output from a lower energy input. The most efficient way to transform energy is by using streams of both north and south magnetic monopoles. The stream of north monopoles always strikes the north monopoles in the energy-producing atom's magnetic line of force, and the stream of south monopoles does the same for the south monopoles.

This creates a circular, polarized wave in the form of a helix that has both north and south magnetic monopoles. The circulating stream of north monopoles attracts the circulating stream of south magnetic monopoles, and this attraction between the monopoles holds the monopoles in the no attract/repel area near the surface of the monopole. This creates a circular, polarized wave that is stable in frequency and able to propagate through the ether for vast distances.

If you were to remove one of the streams of monopoles that an atom's magnetic lines of force were transforming into a circular polarized wave or helix, the stream would then create a circular, polarized wave that is very unstable and unable to propagate very far through the ether. This is the direct result of the missing stream of monopoles, which complements the other monopoles in that it attracts its opposite and therefore holds the helix together.

When a wave of energy is created in this way, it does not have the attraction of the opposite monopoles to keep the waveform stable, resulting in a wave that changes in frequency. This is because the monopoles that created the wave begin to repel each other and can only propagate for a short distance, creating what is known as the Lamb shift.

Alpha particles are also produced in this way. An alpha particle is a north magnetic monopole that has been transformed by an atom's magnetic lines of force without using a stream of south magnetic monopoles. This process gives the alpha particle a positive (north) charge. A beta particle is created the same way, using a stream of south magnetic monopoles instead. This gives the beta particles a negative (south) charge.

Ions are atoms with extra north or south magnetic monopoles attached to their atom's magnetic lines of force. An atom with a positive ion charge simply has an excess amount of north magnetic monopoles attached to its magnetic lines of force. When these extra monopoles are released, they create a stream of north monopoles called positive electricity.

These extra north magnetic monopoles do not increase the magnetic field strength of the atom's lines of force. To be able to increase this magnetic field, you must decrease the atom's temperature, thus increasing the speed of the north and south magnetic monopoles in orbit through its nucleus. Having only an increase in north monopoles does not have any effect on the speed of the atom's magnetic lines of force.

To create ionized gas, you must increase the amount of north or south magnetic monopoles in the atoms' magnetic lines of force. A positive ion has an excess amount of north magnetic monopoles, and a negative ion has an excess amount of south magnetic monopoles. This imbalance is accomplished using concentrated streams of north or south magnetic monopoles (electricity). This stream of monopoles seeks its opposite monopoles in the atoms' magnetic lines of force and attaches itself to its opposite—remember, like poles repel and opposites attract.

Ions are created only if the stream of monopoles is moving slower than the atom's magnetic lines of force, allowing the stream to attach itself to its opposite monopole to form an ion. If this stream of monopoles is moving faster than the magnetic lines of force, it will then strike its own kind of monopole and transform its energy into the atom's magnetic lines of force, creating an electromagnetic wave or helix. This is how plasma is created.

If you were to take two separate metal plates and charge them, the one charged with north magnetic monopoles would create a north, or positive, ion and the one charged with south magnetic monopoles would create a south, or negative, ion. The two metal plates would then have an opposite magnetic charge, so they would attract each other. If these metal plates were allowed to touch, they would discharge their excess magnetic monopoles in the form of electricity. This is due to electricity being created by separating the north and south magnetic monopoles into concentrated streams and is how capacitors are made—by storing these separated north and south magnetic monopoles and then releasing them as electricity.

A battery is a device that uses a chemical reaction that can temporarily store separated north and the south magnetic monopoles, just like a capacitor. Once stored monopoles are separated into concentrated streams, they become electricity. Remember that when two elements combine, they often share magnetic lines of force. This causes release of any excess monopoles that are no longer needed. The greater the speed of this chemical reaction, the greater the voltage the battery produces.

When batteries are cooled to very low temperatures, their atoms' magnetic field becomes stronger. This increase in the atoms' magnetic field slows down the chemical reaction that produces electricity, a phenomenon directly related to the atoms' magnetic lines of force in that it requires more energy to combine the two elements together when they are colder. This slows down the chemical reaction of the battery, producing fewer magnetic monopoles, which results in lower voltage produced by the battery. The opposite occurs if the battery is heated, where the chemical reaction in the battery requires less energy and produces more magnetic monopoles, resulting in a higher voltage output.

Recall that all chemical reactions are affected by changes in temperature. The higher the temperature, the weaker an atom's magnetic lines of force become. Weaker magnetic lines of force enable a chemical reaction to combine elements at a much faster rate. Recall also that conversely, when an atom's magnetic lines of force are cooled, the field becomes much stronger due to less interference by the random movement of north and south magnetic monopoles, called heat. A strong magnetic line of force therefore can slow down, or even stop, a chemical reaction.

For example, reference C of figure 5 shows a small-diameter magnetic line of force. This line of force represents the strong nuclear force. Because of its small magnetic line of force, it is able to transform streams of north and south magnetic monopoles into high-frequency gamma rays. reference C also shows ionic bonding, because element A's and element B's magnetic lines of force are almost the same diameter and so can join together using ions. When these two elements combine, they will require more magnetic monopoles to bind them together. This process reduces the amount of excess monopoles moving at random (heat) and so is a cooling effect.

Figure 5, reference D shows the magnetic lines of force that produce visible light. The large-diameter magnetic line of force creates the color red because of its longer wavelength, and the smaller diameter magnetic line of force creates the color violet because it has a shorter wavelength.

Reference E shows the weak nuclear force. Because of its large diameter, this magnetic line of force is what produces radio waves, which travel at a much slower speed than gamma rays and light waves. Gamma rays travel much faster than light because of their smaller diameter; even though they oscillate at a much higher frequency, their monopoles travel a shorter distance in the same period of time as light. This also illustrates covalent bonding. Element A has the same diameter of magnetic line of force as element B. When these two elements combine, they share the magnetic monopoles that create their lines of force. Because they share magnetic monopoles, the excess monopoles are released as heat. The more magnetic lines of force that combine together, the more heat will be produced.

In reference F, the large, weak magnetic line of force is shown. This magnetic line of force does not produce any electromagnetic waves in the electromagnetic spectrum because its magnetic field is too weak and can be easily removed or combined with other atoms to create the magnetic field in bar magnets. All ferrous atoms have these weak magnetic lines of force like all other elements, though the strength of this magnetic line of force is still dependent on its temperature; the colder it becomes, the stronger its magnetic field becomes. The relative position of this weak magnetic line of force from the atom's nucleus determines if the element is paramagnetic or antiferromagnetic.

All electromagnetic waves are created by the transformation of a concentrated stream of monopoles by an atom's magnetic lines of force. Electromagnetic waves are always circular, polarized, and composed of both north and south magnetic monopoles, creating the particle-wave duality. The only difference in electromagnetic waves transformed by atoms is the waves' frequencies.

6. Forces

Ether

The ether is the medium through which all electromagnetic waves propagate. Nothing can exist without the ether, not even space. This medium occupies all space and is composed of north and south magnetic monopoles that are in north/south pairs and that do not move relative to space. This ether is also responsible for creating inertia, centrifugal force, and momentum. It has the same composition of north and south magnetic monopoles as gravity; however, gravity moves relative to space and the ether does not.

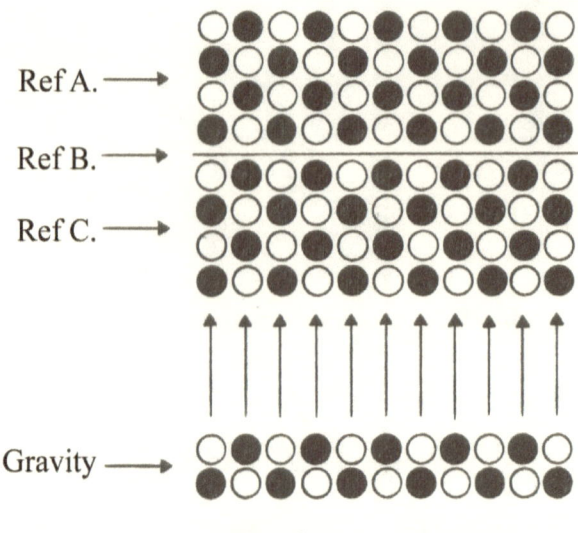

Fig 6 A.

Ether

The ether is shown in figure 6 A. The black dots in this illustration represent north magnetic monopoles, and the white dots represent south magnetic monopoles. This is how the ether would look if you were able to see it. The north and south magnetic monopoles that create the ether are the same monopoles that orbit through atoms' nuclei, creating the atoms' magnetic fields and producing the electromagnetic waves that cover the entire electromagnetic spectrum. Monopoles are the smallest particles that exist, and they are able to pass through everything.

In reference B you can see how all energy propagates through the ether—the ether acts like a conductor for all forms of energy. When gravity enters the ether, it immediately emerges at the other end, because gravity does not propagate like a wave. Instead, gravity acts instantaneously. Imagine a pool table full of balls all lined up in rows, as in reference A. This lineup would represent the ether. When a moving ball of gravity strikes this ether, a ball representing gravity immediately emerges from the other end of this ether.

Light takes approximately eight minutes to travel from the sun to the earth, but gamma rays travel much faster than light, because the frequency of visible light is much lower than that of gamma rays. If you were able to stretch out the monopoles that create a light wave that has traveled the same distance through space as a gamma ray, you would find that the monopoles in the gamma ray traveled a shorter distance than the monopoles in the light wave. The reason for this difference in speed between the light wave and the gamma ray is how the atom transforms energy. When energy is transformed by the atom, it is always circular, polarized, and takes the form of a helix. This is a result of the atom's magnetic lines of force, which transform the energy, being circular.

Gamma rays have a short wavelength because they are created by atoms with small magnetic lines of force. Imagine a wave of energy that is circular polarized. The higher the frequency of this energy, the smaller the diameter the wave will be. This smaller diameter will result in a smaller circumference and a shorter distance that the north and south magnetic monopoles will have to travel, compared to a lower frequency electromagnetic wave. Therefore, the lower the frequency of the electromagnetic wave, the larger its diameter, the greater its circumference, and the greater the distance the magnetic monopoles will have to travel.

Low-frequency electromagnetic waves take more time to travel the same distance than higher frequency waves. This is why you are able to see sunspots on the surface of the sun long before the sun's energy gets to Earth—because light waves have a much shorter wavelength and therefore travel faster than the lower frequency waves like the microwaves that are produced by the increase in sunspot activity.

When a north or south magnetic monopole enters the ether as a wave or as gravity, it propagates as described in the pool ball analogy in figure 6, reference A. Since gravity is not a wave, it emerges at the distant end instantaneously. But for energy that is in the form of a wave, the higher frequencies travel faster since they have less distance to travel than lower frequency waves.

The ether is simply a conductor for all forms of energy; it allows light and energy from distant stars to travel great distances. Without this medium, energy would not be able to propagate, and space itself would not exist.

Gravity

My Theory of Everything defines gravity as an attractive force that attracts all matter and all forms of energy because all things are created from just three particles: north and south magnetic monopoles and particles of matter that represent an element. Gravity is composed of only two particles—the north and south magnetic monopoles. The monopoles that create gravity are in north/south pairs and they are moving in the same direction relative to space. This is the same composition of magnetic monopoles that create the ether; however, the ether does not move relative to space. Gravity can only be produced by a nuclear reaction, the same nuclear reaction that takes place in the cores of all planets and moons, including stars.

During this nuclear reaction, the north and south magnetic monopoles that are in orbit through atoms' nuclei creating the atoms' magnetic lines of force are released as energy. When this release of energy takes place in the center of the earth's core, released monopoles are not able to form a helix. This is the result of the extreme pressure and heat created in the center of the earth's core.

Though these magnetic monopoles cannot form a helix in the earth's core, if you were to move farther away from the center of the earth, away from the extreme heat and pressure, the atoms' magnetic lines of force would be able to transform streams of monopoles into helices, which are circular, polarized electromagnetic waves. Due to the extreme heat and pressure in the center of the earth's core, the magnetic monopoles that would create electromagnetic waves cannot oscillate. This forces the magnetic monopoles to move in north/south pairs and in the same direction—out from the center of the earth's core.

These are the north and south magnetic monopoles that create gravity. They can only move out from the center of the earth's core. As they do so, they attract all matter and all electromagnetic waves because all matter is bound together by the north and south magnetic monopoles and all electromagnetic waves are created by magnetic monopoles. This will cause all things to be pulled toward the center of the earth, creating a sphere.

Gravity is only a weak attractive force because it can only attract a small area of atoms' magnetic lines of force. The more magnetic lines of force that orbit through an atom's nucleus, the heaver the element and the higher that element's atomic number. Since gravity is created by the north and south magnetic monopoles that are moving in the same direction relative to space, these monopoles attract the north and south magnetic monopoles that are in orbit through the atoms' nuclei. Gravity is a weak force due to the movement of these magnetic monopoles through the atoms' nuclei because the north magnetic monopoles are moving in a clockwise direction and the south magnetic monopoles are moving in a counterclockwise direction.

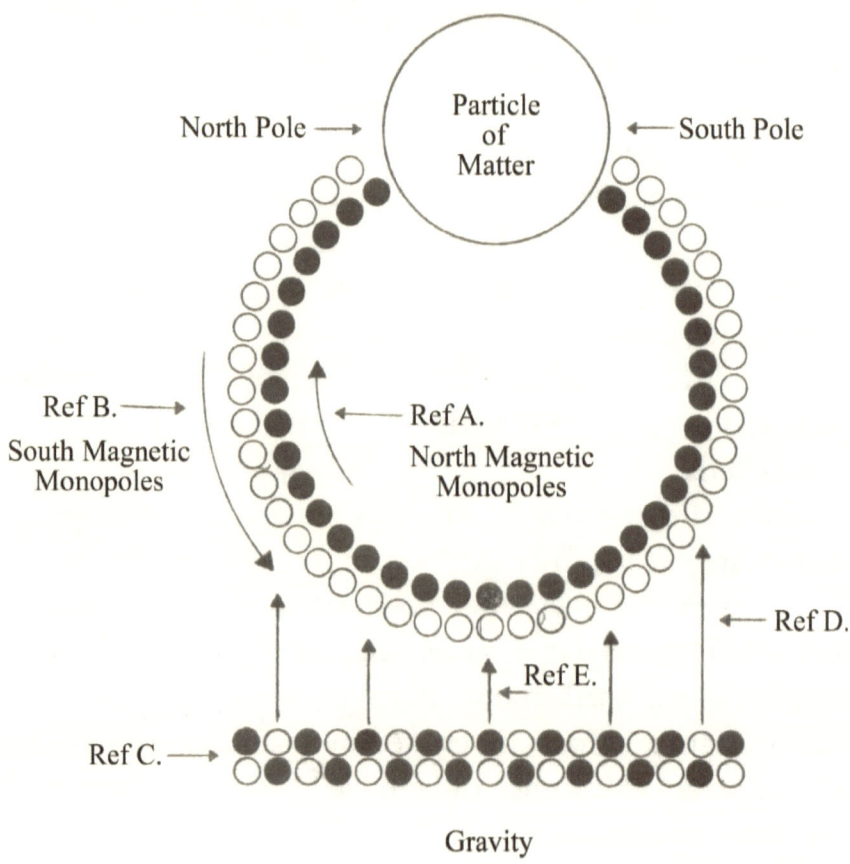

Fig 6 B.

Gravity

Figure 6 B, reference A shows the north magnetic monopoles moving in a clockwise direction through an atom's nucleus. They also make up one-half of the magnetic line of force.

In reference B, see that the other half of a magnetic line of force is created by south magnetic monopoles. These monopoles are moving in a counterclockwise direction through the atom's nucleus. Because these north and south magnetic monopoles attract each other, they continue this perpetual motion, creating a magnetic line of force and producing a magnetic dipole, just like a bar magnet.

Reference C represents the north and south magnetic monopoles that create gravity. These monopoles are all moving in the same direction relative to space. This is also the same composition of magnetic monopoles that create the ether.

In reference D, the north and south magnetic monopoles that create gravity also attract the north and south magnetic monopoles that are in orbit through the atom's nucleus. Since the monopoles that create the atom's magnetic lines of force are moving relative to the magnetic monopoles that create gravity, they cancel the attraction between them. This movement of monopoles is what makes gravity a weak force. When a north magnetic monopole is moving in the same direction as a south magnetic monopole, their attraction is canceled out. When a south magnetic monopole is moving toward a north magnetic monopole, the attraction between them is again canceled out.

For example, if you were to jump out of an airplane, while you were falling down to earth, you would become weightless. Because you are moving toward the source of gravity, just like the north magnetic monopole that is moving toward the south magnetic monopole, the gravity's attraction is canceled out. However, when a south magnetic monopole from an atom's magnetic line of force moves in the same direction as the north magnetic monopole of gravity, they attract each other. This does not have any effect on the atom because the monopoles in the atom's magnetic line of force are also moving in the same direction.

Reference E shows the only area that north and south magnetic monopoles of gravity are able to attract the monopoles that create the atom's magnetic lines of force. When the magnetic monopoles are in this position, they are moving exactly horizontal to the north and south monopoles of gravity. Only at this point is gravity able to attract the atom's magnetic lines of force. This is why gravity is the weakest form of energy.

Gravity also attracts all electromagnetic waves in the same way it attracts the atom, because electromagnetic waves are produced by the atom's magnetic lines of force and form a helix.

Centrifugal Force

Many scientists believe that centrifugal force is a fictitious force. It is defined as the force that tends to make rotating bodies move away from the center of rotation. In fact, it is an attractive force created by the north and south magnetic monopoles that create the ether.

Centrifugal force is simply another name for gravity. They are both created by north and south magnetic monopoles that are interacting with the magnetic monopoles in atoms' magnetic lines of force.

Centrifugal force is the result of an object traveling in a circle. Such objects behave as if they are experiencing an outward force, the strength of which depends on the mass of the object, the speed of rotation, and the distance from the center. The more massive the object, the greater the force that is felt. Likewise, the greater the speed of the object, the greater the force, and the greater the distance from the center, the greater the force.

This attractive force is created by the north and south magnetic monopoles that create the ether. These magnetic monopoles interact with the north and south monopoles that create the magnetic lines of force orbiting through atoms' nuclei giving the atom its magnetic field. It's this magnetic field that binds atoms together to create matter. When an object rotates on its axis, the north and south magnetic monopoles that create the atoms' magnetic lines of force move past the north and south magnetic monopoles that create the ether. These atomic magnetic monopoles then attract the monopoles that create the ether.

Due to angular momentum, the monopoles that create the atoms' magnetic lines of force will always be separated from the magnetic monopoles that create the ether. This separation of monopoles creates the same effect as gravity. When an object is stationary, the north and south magnetic monopoles that create gravity are attracting the atoms' magnetic field.

You may say that gravity, centrifugal force, inertia, and momentum are all created by the same thing. When an object moves relative to the ether or space, the magnetic monopoles that create the atoms' magnetic lines of force interact with the magnetic monopoles that create the ether.

North and south magnetic monopoles always seek their opposites, meaning a north monopole always attracts a south monopole. When this north/south magnetic monopole combination is separated, it creates resistance that lasts while they are kept separated, and this creates the effect known as inertia and momentum, or centrifugal force.

When an object is at rest relative to space, the north and south magnetic monopoles that create the ether are no longer interacting with the magnetic monopoles that create the atoms' magnetic lines of force, and therefore are not creating centrifugal force, momentum, or inertia. When an object is moving relative to space, as its velocity increases, the resistance between the separating north/south magnetic monopoles also increases. It's this increase in resistance that causes an increase in centrifugal force. Remember, inertia, momentum, centrifugal force, and gravity are all different names for the same thing. They are all created by the monopoles of the ether interacting with the monopoles in the atoms' magnetic lines of force.

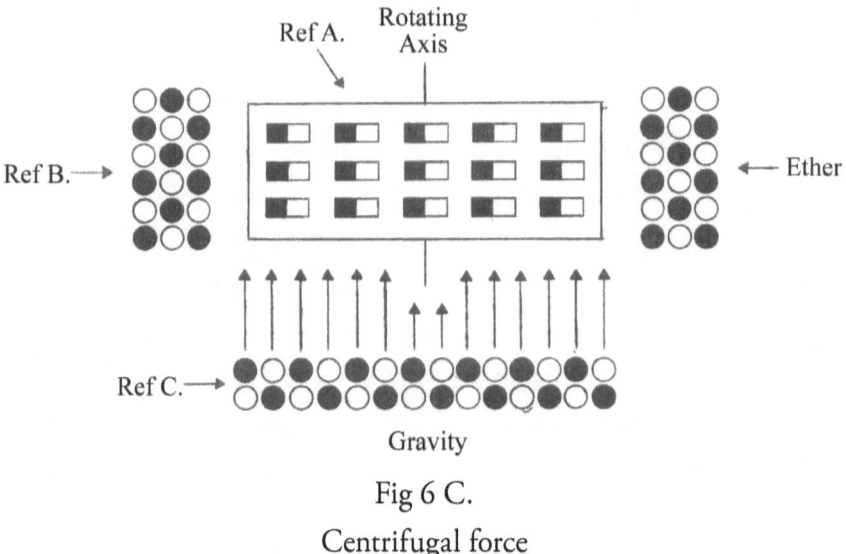

Fig 6 C.

Centrifugal force

In figure 6 C, reference A shows the atoms in a rotating mass as small magnets, because each atom is a magnetic dipole with its north and south magnetic monopoles orbiting through its nucleus in opposite directions, creating a magnetic line of force and producing the atoms' magnetic fields. These monopoles from the atoms' magnetic lines of force attract the same north and south magnetic monopoles that create gravity. When this object is rotating, the magnetic monopoles that create the atoms' magnetic lines of force will be always separating from the magnetic monopoles that create the ether, creating the same effect as gravity—an attractive force.

Reference B shows north and south magnetic monopoles that are not moving relative to space—the ether that occupies all space and is how all energy propagates. The north and south magnetic monopoles that create the ether are responsible for creating inertia, momentum, and centrifugal force.

Reference C illustrates how gravity is created by north and south magnetic monopoles that are moving alongside each other and in the same direction. It's this movement of magnetic monopoles that attracts all atoms since each atom has a magnetic field created by the same north and south magnetic monopoles that create gravity. The force of gravity can only attract a small area of the atom's magnetic lines force because the magnetic monopoles that create the atom's magnetic lines of force are perpetually moving in opposite directions.

When the centrifugal force on a spinning object equals or exceeds the force of gravity that is pulling the object down, the object becomes stable relative to space. In other words, it will no longer wobble like a top that is slowing down. This is because centrifugal force acts just like gravity; however, it pulls on the circumference of the rotating object whereas gravity pulls on the bottom of the object.

7. Space

Black Holes

Black holes are stars that have exhausted all their energy. They are composed of particles of matter that represent many different elements from the periodic table.

The sun, like all stars, is a large nuclear reactor. Stars produce energy in all wavelengths of the electromagnetic spectrum, from radio waves to gamma rays, including gravity. This energy is created from the north and south magnetic monopoles that are in orbit through the atoms' nuclei. When these north and south magnetic monopoles begin to run out, the star's gravity weakens.

The weakening of the star's gravity allows the star to expand, further weakening its gravity and energy.

It's this force of gravity that keeps the sun from expanding, keeping pressure on its core by attracting the north and south magnetic monopoles on its surface.

Once the weakening of the star's gravity reaches a critical point, the star's nuclear reaction becomes unstable. Without the pressure created from the force of gravity, the star explodes, creating a supernova. This uncontrolled nuclear reaction or supernova uses up all of the remaining north and south magnetic monopoles that are in orbit through the atoms' nuclei. These monopoles are all released at once. After this massive release of energy, all of the star's atoms no longer have any north and south magnetic monopoles orbiting through their nuclei. Now all that is left from the star are the particles of matter that represent the many different elements. Only stars with enough initial mass can become black holes.

The elements in a black hole had once made up the stars' atoms' nuclei. Without the north and south magnetic monopoles, there cannot be any magnetic lines of force orbiting through the atoms' nuclei, and without these lines of force, there cannot be a magnetic field to bind the atoms together. Now these particles of matter that represent elements must begin to attract the same north and south magnetic monopoles they had released as energy so these particles of matter or elements can become atoms again. Because there are so many particles of matter that are attracting so many north and south magnetic monopoles, the particles of matter begin to collapse onto themselves to form one solid mass containing only particles of matter.

This solid mass of matter becomes very dense. Since the particles of matter no longer have north and south magnetic monopoles orbiting through their nuclei, they cannot create the magnetic lines of force that produce magnetic dipoles.

Without this magnetic field, atoms cannot be created, so these particles of matter are held together in one solid mass by the pressure of the north and south magnetic monopoles that are on the surface of this mass of particles of matter. It creates the same effect as gravity in that it keeps pressure on the mass of matter. This is what keeps the particles of matter from exploding.

Black holes are the exact opposite of stars in that they draw energy into their cores of matter. This energy is in the form of north and south magnetic monopoles, the same monopoles that were in orbit through the nuclei of the atoms within and that gave the atoms their magnetic fields. Black holes are black because they attract the north and south magnetic monopoles that create light. Finally, when the particles of matter in a black hole have their north and south magnetic monopoles back, they will become atoms again.

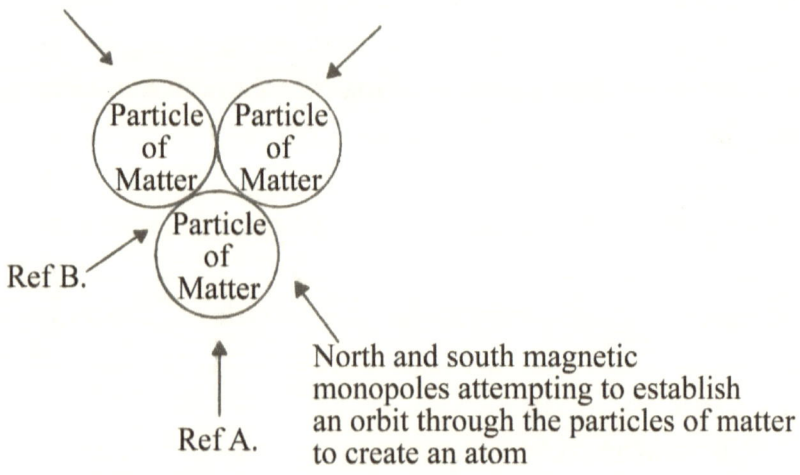

North and south magnetic monopoles attempting to establish an orbit through the particles of matter to create an atom

Fig 7 A.

A black hole

Figure 7 A, reference A shows what a black hole would look like if you were able to see it. Only three particles of matter are shown in this illustration; however, black holes contain many particles of matter. These particles of matter do not have a magnetic field because they no longer have any north and south magnetic monopoles orbiting through the atoms' nuclei. This is because those monopoles have all been released during a supernova. Now these particles of matter will begin to attract these north and south magnetic monopoles in order to create magnetic lines of force so they can become atoms again.

In reference B, after a supernova, when all the atoms have used up all their energy and all of the north and south magnetic monopoles are gone, the particles of matter begin to collapse into one large mass of matter containing only elements. This mass of particles will be under tremendous pressure, created by the north and south magnetic monopoles that are trying to get back into orbit through the nuclei of the particles of matter so they can become atoms again. The pressure keeping the particles of matter in one large mass is created by the magnetic monopoles on the surface of the black hole. The particles of matter on the surface have north and south magnetic monopoles orbiting through their nuclei, creating magnetic lines of force. It's these lines of force that keep pressure on these particles of matter inside, because the north and south magnetic monopoles are being drawn into the center of the black hole by the particles of matter within. These particles do not yet have their north and south magnetic monopoles back, so nothing orbits through their nuclei.

This movement of monopoles is a slow process. The great pressure created by the atoms on the surface of the black hole keeps the particles of matter in the center of the black hole from expanding and allowing the north and south magnetic monopoles to return. When the particles of matter in the center of a black hole attract these north and south monopoles, the monopoles must move in one direction alongside each other, and in the same direction. This is the same composition of monopoles as gravity, but it is created in the opposite direction, which is why black holes create strong gravity fields.

Eddy Currents

Eddy currents are created by the constant reversal of an atom's magnetic poles, which is the result of an alternating magnetic field. Specifically, eddy currents are the resistance to this change of magnetic poles. An atom is a magnetic dipole in that it has both north and south magnetic poles. These magnetic dipoles are created by the perpetual motion of north and south magnetic monopoles, which create a magnetic line of force, which then produces a magnetic dipole. It's this magnetic dipole, called an atom, that is responsible for creating eddy currents.

When atoms are in the presence of a magnetic field, their magnetic poles align themselves to the magnetic lines of force from this external magnetic field. The atom, as a small magnetic dipole, acts the same way as a compass needle would act in the same magnetic field since a compass needle is also a magnetic dipole.

Atoms are just small magnets, so one would think that they would stick to a large bar magnet. However, that's not the case. For an atom to be able to stick to a bar magnet, its magnetic lines of force must curve immediately as they pass through the atom's nucleus. That would attract the poles of the bar magnet to the poles of the atom, bringing the atom closer to the bar magnet. However, the bar magnet is so much larger than the atom that its magnetic lines of force always appear to be straight lines, so they pass right through the atom's nucleus. The only thing that the atoms do is orientate their magnetic poles to point to the bar magnet's magnetic poles, just like a compass needle would. This is the same reason why a compass needle, also a magnetic dipole, always points to the earth's magnetic poles.

The same thing happens to the atoms that create matter. For example, aluminum is made from atoms that resist change in magnetic polarity. When in the presence of a changing (or alternating) magnetic field, the atoms that create aluminum resist this change in magnetic polarity and then repel from the alternating magnetic field.

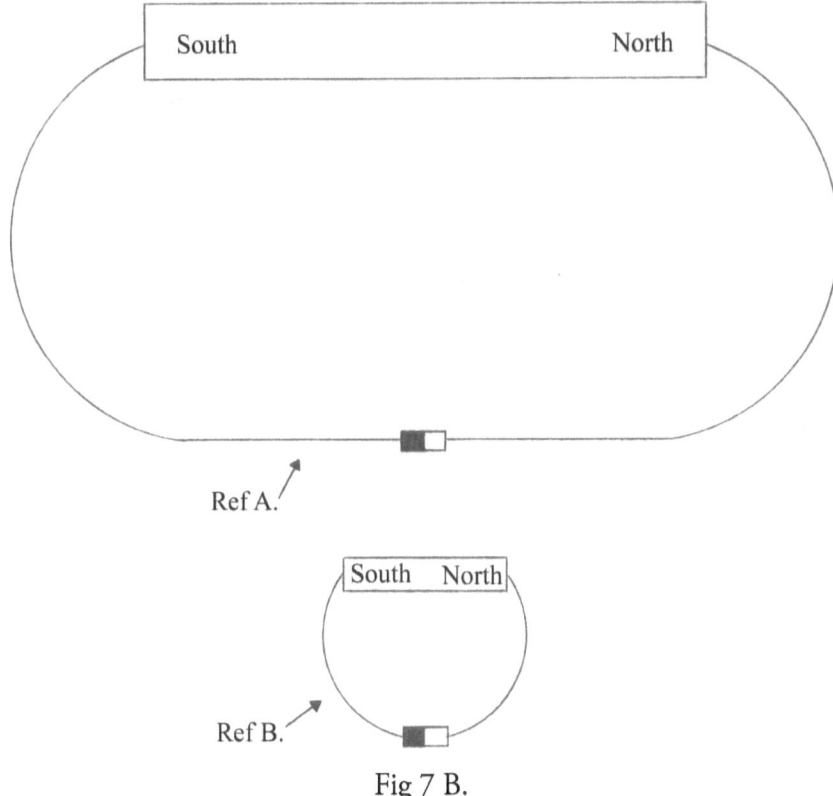

Fig 7 B.

Large- and small-scale magnetic lines of force

Figure 7 B, reference A shows how an atom orients its magnetic poles to the lines of force from a bar magnet, just like a compass needle points to the earth's magnetic poles. The atom, a small magnetic dipole, is not attracted to the bar magnet because its magnetic lines of force do not curve immediately after they pass through the atom's core.

In reference B, notice that if the bar magnet were much smaller, its lines of force would curve immediately after they have passed through the atom's core. The atom would then become attracted to the bar magnet. The north and south magnetic monopoles that create the bar magnet's lines of force would then pull the atom toward the bar magnet's north and south poles.

Eddy currents, the reversal of an atom's magnetic poles, can also be used to heat matter. This is accomplished using a high-frequency alternating magnetic field. Such a magnetic field causes the atom's magnetic poles to change direction rapidly. This constant change in the atom's magnetic field slows down the perpetual motion that creates the atom's magnetic lines of force. Heat, the random movement of monopoles, also slows that perpetual motion, having the same effect as high-frequency eddy currents. This weakens the magnetic bond between the atoms. If this magnetic bond becomes too weak, the matter will change from a solid to a liquid.

Earth and the Moon

The same magnetic monopoles that make up all forms of energy and create gravity and magnetism also play a role in how the planets and the moon move in the solar system.

The moon orbits the earth approximately every 27.5 days and always has the same side facing the earth. It also has a nuclear reaction taking place in its core, which is what creates the moon's gravity. Matter that is drawn toward the center of an object by the force of gravity always creates a sphere; mass does not have the properties of gravity. If it did, the planets would not be spheres.

The moon, just like the planets, has a magnetic field. Its magnetic north pole is pointing in the same direction as all the other planets, including stars. This magnetic field is created by the combined weak, outer magnetic lines of force that come from the ferrous atoms that the moon and planets are made of. It's this magnetic field that keeps the moon from striking the earth despite the force of gravity from the moon and earth pulling on each other, bringing them closer together. The moon's magnetic field pushes it farther away from the earth's magnetic field because the moon's magnetic north pole is on the same side as the earth's magnetic north pole. Because of this alignment, the two bodies will always repel each other.

As the force of gravity begins to pull the moon and the earth closer together, their magnetic poles begin to repel each other. This tug-of-war continues until the two forces, gravity and magnetism, balance each other. The moon revolves around the earth every 27.5 days, and in the same direction as the earth rotates, because of the earth's gravitational field, which originates in the center of the earth's core and rotates along with the earth.

This rotating gravitational field creates a horizontal as well as a vertical component of gravity. The vertical component of gravity is what pulls the moon closer to the earth, and the same vertical component of gravity is what keeps objects on the surface of the earth. The horizontal component of gravity is what causes the moon to revolve around the earth.

The horizontal component is created by the earth rotating on its axis. The part of the horizontal component of the earth's gravity that is moving away from the moon also attracts the moon, pulling it in the same direction that the earth and its gravitational field are rotating. Meanwhile, the horizontal component of gravity that advances toward the moon does not have any attraction because the source of this horizontal component is moving closer to the moon, canceling the force of gravity. This is why the moon revolves around the earth in the same direction that the earth rotates.

The moon always keeps the same side facing the earth. This is because it does not rotate on its axis as it revolves around the earth. This is the result of the combined force of gravity from the earth and moon, which are always pulling on each other. This attractive force is much stronger than the rotating force of gravity from the earth that would otherwise cause the moon to rotate on its axis. If the moon were farther away from the earth, it would begin to rotate on its axis because the combined forces of gravity from the moon and earth would be weaker than the earth's rotating force of gravity.

The earth and all of the planets revolve around the sun because of the sun's gravity. When the sun rotates on its axis, the force of gravity that is created by the nuclear reaction in its core also rotates at the same rate as the sun, and it's this rotating gravitational field that causes the earth and all the other planets to revolve around the sun in the same direction as the sun rotates. If the sun's gravitational field were not rotating, there would not be any horizontal component of gravity to cause the planets to revolve or rotate.

Imagine you are moving away from the earth's gravity. You would feel the force of gravity increase. The same thing happens as the horizontal component of gravity that emanates from the sun that is moving away from the planet; the effect is the same. When the horizontal component of the sun's rotating force of gravity advances toward the planet, it causes the advancing horizontal component of gravity to become weaker.

If you were to jump off a tall building, while you were falling, you would also be moving closer to the source of gravity. It's this movement that makes you feel weightless. The same thing is happening to the planet. Now you have a stronger force of gravity on the left side of the planet than you have on the right side (see figure 7 D). This imbalance in the force of gravity causes the planet to revolve around the sun and also to rotate on its axis. The farther a planet is from the sun, the slower it revolves around the sun and the faster it rotates on its axis. This is all the result of the sun's horizontal component of gravity.

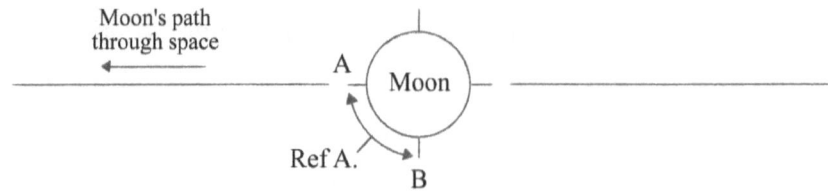

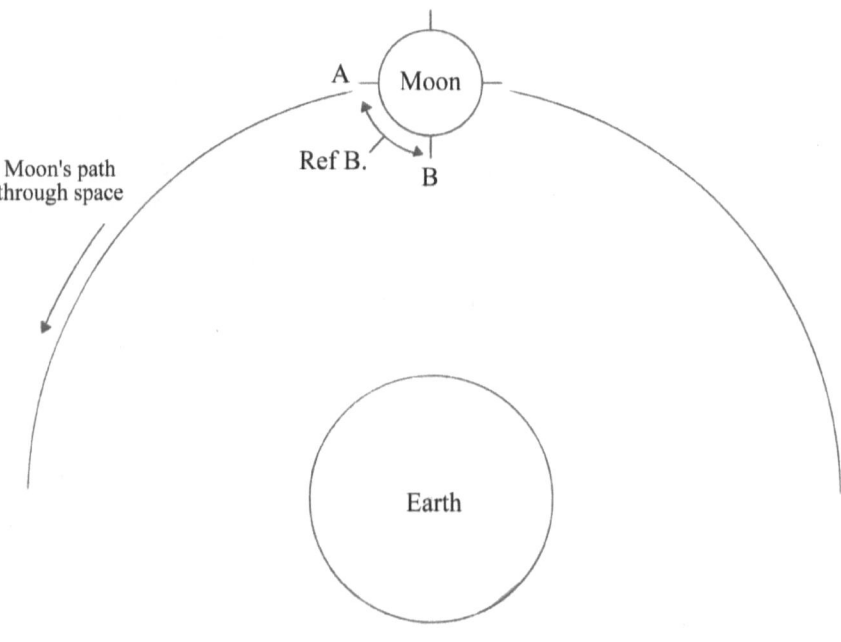

Fig 7 C.

The moon's orbit

In figure 7 C, reference A, Area A is the direction that the moon moves through space and Area B is the side of the moon that is always facing you. The angle between Areas A and B does not change because the moon does not rotate on its axis. When the moon's axis moves through space in a circle with the earth in the center, the moon always has the same side facing the earth.

Reference B, Area A shows the direction that the moon is moving through space, while Area B is the side on the moon that you always see. This is because the moon does not rotate on its axis. Since Area B does not cross the path that the moon's axis takes as it moves through space, the only difference between reference A and reference B is the path the moon is shown taking through space.

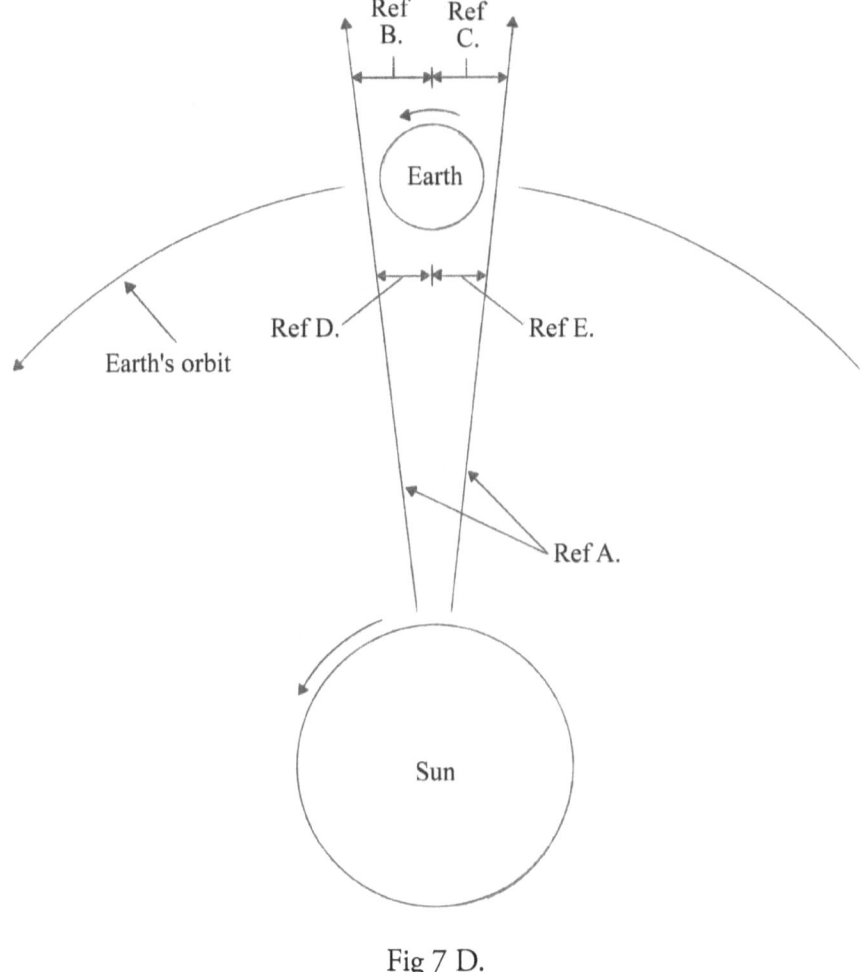

Fig 7 D.
Earth's orbit

In figure 7 D, reference A, the two vertical lines in this diagram represent the horizontal component of gravity, which is moving in a counterclockwise direction. This horizontal component of gravity is created by the sun rotating on its axis. The horizontal component of gravity that is advancing toward earth has less attraction to the earth because the force of gravity is advancing closer to the earth. The horizontal component of gravity that is moving away from the earth has a greater attraction to the earth because it is moving away from the earth. Remember, when an object is falling to the earth, or toward any source of gravity, it becomes weightless. The opposite occurs when an object moves away from the source of gravity. So in this diagram, the left side of the earth has a stronger attraction to the sun's rotating gravitational field than the right side, which is what causes the earth and all of the other planets to revolve around the sun. It's also the same rotating gravitational field that causes the earth and the planets to rotate on their axes.

The sun's horizontal component of gravity is greater at reference B than it is at reference D. This is the result of the distance the horizontal component of gravity has traveled during the same period of time, which makes the force of gravity greater on the far side of the earth than the near side. As the sun's horizontal component of gravity approaches the earth, the force of gravity becomes less at reference C than at reference E because of the distance traveled. This is why the earth and all other planets rotate on their axes and why the planets revolve slower and rotate faster the farther they are from the sun.

The Seasons

The four seasons are the direct result of the earth's tilt. Because the earth is tilted on its axis, the northern and southern hemispheres do not receive an equal amount of the sun's energy at all times. It's this imbalance that creates the seasons, but why is the earth tilted on its axis?

The earth's magnetic field is the reason for this. The earth, like all other moons and planets, has a magnetic field. This magnetic field is a dipole in that it has a magnetic north and south pole, just like a bar magnet. All of the planets and moons have their magnetic north poles facing the same direction as the sun's magnetic north pole. This magnetic orientation creates a repulsive force between the planets and moons; they all repel each other. These planets and their moons also have nuclear reactions going on in their cores, creating gravity, an attractive force. Gravity attracts all the moons and planets, including the sun, and brings them closer together. Because their magnetic fields all have their north poles facing the same direction, they begin to repel or push away from each other. This tug-of-war continues until the forces of gravity and the magnetic fields become equal.

This balance of magnetism and gravity is what holds the planets and their moons in orbit around the sun, as explained earlier in this chapter. However, not all things are equal, and the north and south magnetic monopoles that create the magnetic lines of force (which in turn create magnetic dipoles) are not equal. We know this because we can see the results of this magnetic imbalance. As the earth's magnetic poles get closer to the sun's magnetic poles, the north pole repels farther away than its south pole. This is caused by the difference in strengths of its north and south magnetic poles.

The south magnetic monopoles are not as strong as the north magnetic monopoles—they are very close, but not exactly equal in strength.

Imagine you had only two magnetic monopoles, one north and one south. They would appear to be equal in strength. Now increase both types in equal number. Remember that magnetic monopoles are quantified in that they can combine together and become stronger. Having a large-but-equal amount of north and south magnetic monopoles, their difference in strength soon becomes apparent.

When these magnetic monopoles are in perpetual motion through a planet's or a moon's core, they create a magnetic line of force, which produces a magnetic dipole. This magnetic dipole has a magnetic north pole that is stronger than its magnetic south pole. For example, recall that an atom is just a small magnetic dipole. The difference in strength between the north and south pole is very small—so small that it usually goes unnoticed. However, increase the size of a magnetic dipole (for example, the planet Earth) and now you have a much greater difference in strength between the north and south poles due to the large quantity of magnetic monopoles involved.

The difference in strength is dependent upon the size of the magnetic dipole. In other words, the stronger the magnetic field, the more magnetic monopoles it took to create the magnetic line of force and the greater the difference in strength between the north and south poles will be. It's this difference in strength that causes the earth to tilt on its axis, which in turn is what creates a change in the seasons. Finally, this difference in strength between the magnetic poles is also responsible for the earth having an elliptical orbit around the sun. The same holds true for all the other planets and moons.

But how do gravity and magnetism work together to create an elliptical orbit?

The difference in strength between the north and south pole causes the planet to tilt on its axis, and therefore its magnetic north pole moves farther away from the sun than its magnetic south pole. When you factor in the force of gravity, this force pulls the planets back together. Since the planets are very large, they also create a large amount of inertia, and it's the combination of these three forces—gravity, inertia, and magnetism—that interact with each other and create an oscillation between the planets and the sun.

The north pole of the planet repels away from the north pole of the sun more so than the south poles do. This is due to the difference in strength between the north and south magnetic poles. When you add the force of gravity, an attractive force, the sun and the planet remain pulled close together. These two forces, gravity and magnetism, would normally equalize, and the planet would stay the same distance away from the sun. However, you must factor in the force of inertia, which interacts with gravity and magnetism and creates an elliptical orbit. When the north pole of the planet repels away from the sun's north pole, it must first overcome the force of inertia, which was created by gravity pulling the two bodies closer together. Once the planet begins to move farther away, weakening the repulsive force of the north poles, the force of gravity must now overcome the force of inertia created by the magnetic poles repelling each other. This again brings the planet back closer to the sun. Now that the planet is moving closer to the sun, it continues to move closer until the repulsive force of the planet's magnetic poles overcomes the force of inertia created by gravity. This cycle continues, causing the planet to tilt on its axis and giving it an elliptical orbit around the sun.

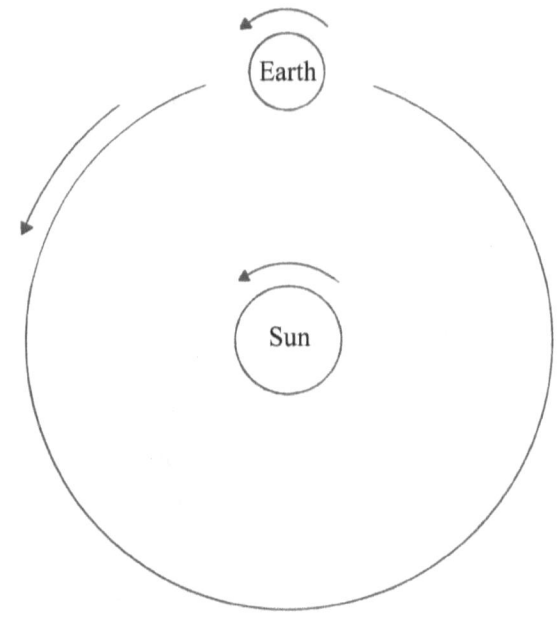

Apogee   Perigee

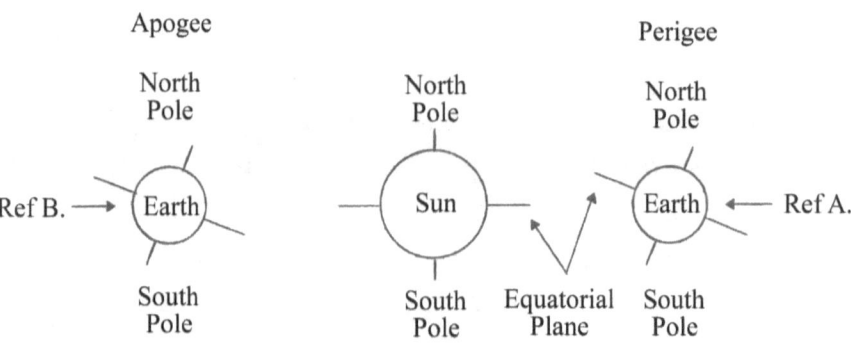

Fig 7 E.

Earth's seasons

Figure 7 E, reference A shows what occurs when the north pole is stronger than the south pole. When the earth is at its perigee, the north pole is repelled farther away from the sun than the south pole due to the difference in strength between the north and south poles. This difference creates the tilt in the earth's axis, which is what creates the changing seasons. After overcoming the force of inertia created by gravity pulling the earth closer to the sun, the earth begins to move farther away from the sun because the magnetic poles repel each other.

In reference B, after the magnetic poles have pushed the earth away from the sun, the momentum created by the magnetic poles repelling each other now moves the earth beyond the point where the earth's magnetic field and its gravity field are equal. Now that the earth is at its apogee, the sun's and earth's gravity, an attractive force, has to overcome the inertia created by the magnetic poles repelling each other. Gravity now begins to bring the two bodies back closer together. When the earth reaches its perigee, the earth and the sun's north poles will again repel each other more than the south poles, causing the earth to tilt on its axis. It must now overcome the force of inertia created by the force of gravity, which pulls the earth closer to the sun. This cycle between the forces of gravity and magnetism continues indefinitely.

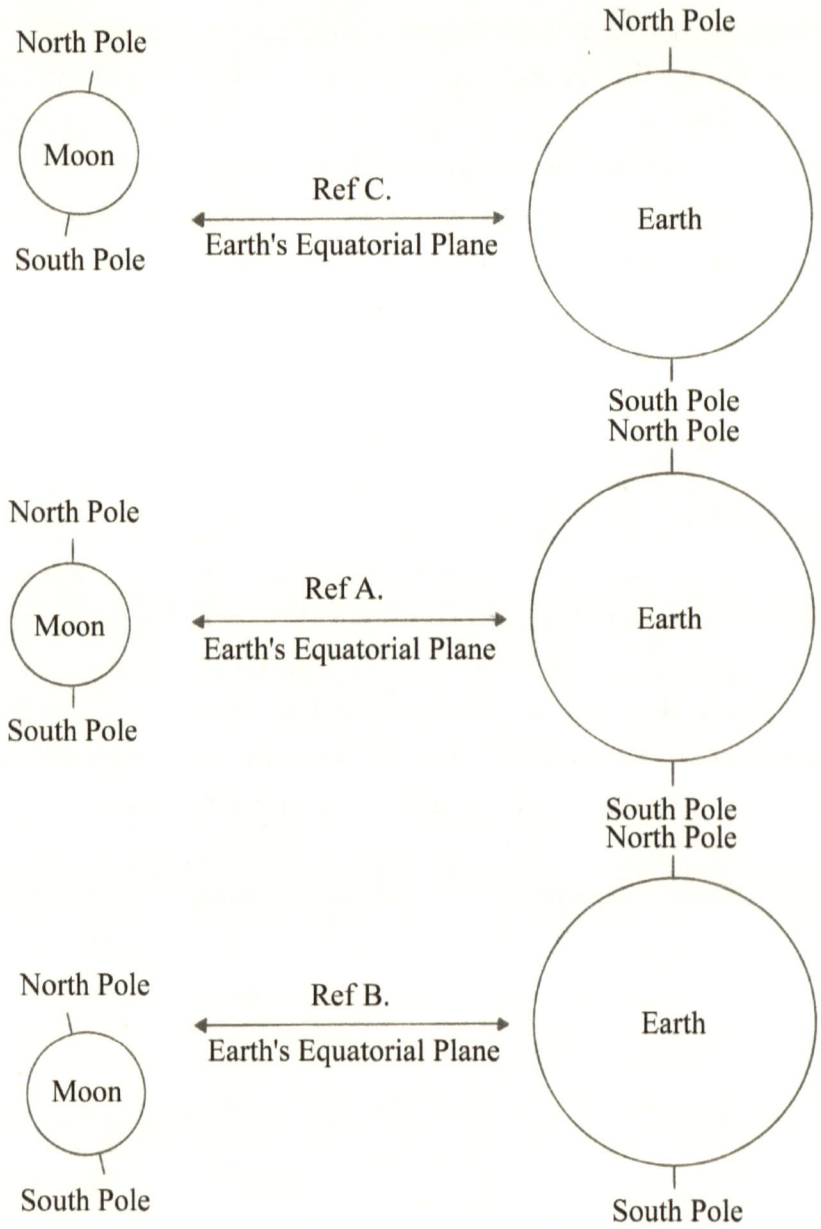

Fig 7 F.
The moon's tilt

Reference A displays how the earth and moon have their north poles on the same side, just like all other planets do. This creates a tilt in the moon's axis, just like the planets. However, the force of gravity is much stronger between the earth and moon then other planets, because they are much closer to each other. This increase in the force of gravity prevents the moon from moving very far from the earth and creates only a slight elliptical orbit around the earth.

As seen in reference B, because the moon is unable to move very far from the earth (due to the strong force of gravity) and because their north poles are stronger than their south poles, when the moon's north magnetic pole begins to repel from the earth's north magnetic pole, it creates a slight tilt in the moon's axis and pushes the moon below the earth's equatorial plane. Now that the moon is below the earth's equatorial plane, the south poles have been brought closer together by the momentum created by repulsive force of the earth's and the moon's north poles. The moon's south pole now begins to repel the earth's south pole. After this repulsive force overcomes the inertia that was created by the north poles repelling each other, they begin to move closer to the north magnetic pole, passing the earth's equatorial plane. This once again brings the earth's and the moon's north poles much closer together.

In reference C, now the moon has moved past the earth's equatorial plane, it continues to move in this direction due to the force of momentum that was created by the south poles repelling each other. This continues until the north poles begin to repel each other, with the moon again having to overcome the force of inertia. Because the moon has moved past the earth's equatorial plane, it creates an orbit that moves above and below the earth's equatorial plane and is slightly elliptical.

<u>*Aurora Borealis*</u>

The aurora borealis is commonly known as the northern lights. This heavenly phenomenon is created by north and south magnetic monopoles in the form of electromagnetic waves.

The aurora borealis is created in the same way as all other forms of energy—by streams of magnetic monopoles that are transformed into electromagnetic waves by atoms' magnetic lines of force. In the case of the aurora borealis, the streams of magnetic monopoles are in the form of electromagnetic waves that were created by the nuclear reaction that takes place within the sun.

This supply of magnetic monopoles that is coming from the sun is transformed by the atoms in the earth's atmosphere, which is composed of approximately 79 percent nitrogen and 21 percent oxygen. The transformation of energy begins by the streams of magnetic monopoles in the form of electromagnetic waves that strike the magnetic lines of force that are in orbit through the atoms' nuclei that create the nitrogen and oxygen in the earth's atmosphere. However, before this transformation can begin, the magnetic monopoles in the form of electromagnetic waves must be able to reach the earth's atmosphere.

The earth is a large magnetic dipole because it has both a north and a south magnetic pole. These magnetic poles are created by north and south magnetic monopoles that are in orbit through the earth's core and that create a magnetic line of force. These monopoles from the sun must first pass through the earth's magnetosphere. The lower frequency electromagnetic waves are composed of slower moving magnetic monopoles, which will not be able to pass through the earth's magnetosphere. This is a result of the monopoles that create the earth's magnetic field, which are moving much faster than the monopoles that created the lower frequency electromagnetic waves.

Before a magnetic monopole can pass through the earth's magnetosphere, it must be moving at the same speed as or faster than the magnetic monopoles that created the earth's magnetic lines of force. High-frequency electromagnetic waves are able to pass through the earth's magnetic field because their magnetic monopoles are moving faster than the earth's monopoles, which allows them to reach the atoms in the earth's atmosphere. However, this does not mean that the magnetic monopoles that reached the earth's atmosphere are moving faster than the magnetic monopoles that are in orbit through the nuclei of the atoms that make up the earth's atmosphere. Before energy can be transformed by the atom's magnetic lines of force, the stream of monopoles must be moving at the same speed or faster than the monopoles in the magnetic line of force that will transform this energy. The only way to increase the speed of these magnetic monopoles is to increase the frequency of the electromagnetic waves that are coming from the sun.

Our sun has an eleven-year cycle. During the peak of this cycle, the sun experiences more solar activity, or sunspots, than usual. These sunspots create massive amounts of magnetic monopoles in the form of gamma and x-rays, which are the highest frequency electromagnetic waves that the sun is able to create. These magnetic monopoles that create gamma and x-rays are moving much faster than the monopoles that create visible light. When these fast-moving monopoles strike the atoms in the earth's atmosphere, they are able to transform into visible light. The color of this light is dependent upon the magnetic line of force that is struck; the smaller the diameter of magnetic line of force, the higher the frequency, which also means a change in color. The color red is a lower frequency of electromagnetic wave than the color blue. Thus, an electromagnetic wave requires a stream of faster moving magnetic monopoles to produce blue light than are required to produce red light.

An increase in solar activity is not only responsible for the aurora borealis, but is also responsible for power outages. When large amounts of magnetic monopoles strike the earth due to an increase in solar activity, they induce magnetic currents in the earth's power grid. Remember, electricity is created by separating north and south magnetic monopoles into concentrated streams. Large amounts of outside magnetic monopoles striking the earth act the same way as an electric generator, but on a much smaller scale. Electricity is generated by a magnet moving past the atoms' magnetic lines of force in a coil of wire, which separates the magnetic monopoles in the coil of wire into concentrated streams. The same thing happens with the power grid. This is caused by the large amount of magnetic monopoles striking the earth.

When a north or south magnetic monopole moves past an atom's magnetic line of force in a wire, it attracts its opposite monopole. This opposite monopole moves to the end of the wire, while the monopole from the atom's magnetic line of force is replaced by one of the many surplus magnetic monopoles in the ether. Since you have a constant stream of north and south magnetic monopoles moving past the magnetic lines of force created by the atoms that make up the wires in the power grid, this flow of magnetic monopoles produces a direct current (DC) voltage in the power lines. However, in an electric generator, an alternating current

(AC) voltage is produced because the generator uses a large magnet to separate the magnetic monopoles into concentrated streams. In the case of the power grid, electricity is being generated by many small magnets that are moving in both directions, producing a DC voltage in the power grid.

Because the power grid was designed for AC voltage, it does not allow DC voltage to pass through its transformers in the substation. This DC voltage instead creates a magnetic field in the primary of the transformer. This prevents the AC voltage from passing through to the transformer's secondary, thereby stopping all AC voltage in the power grid and creating a wide-area blackout.

High-frequency (HF) Radio wave Propagation

Radio waves are electromagnetic and are created from electricity. Electromagnetic waves, created by the nuclear reaction that takes place in the sun, are all circularly polarized. This type of polarization combined with a long wavelength makes radio waves have difficulty in passing through an area where monopoles are moving in opposite directions, like the earth's magnetosphere. Imagine you are on a river and you have to paddle your boat upstream for a long distance. Just like electromagnetic waves that have a long wavelength have to move against the magnetic monopoles in the earth's magnetosphere, you would not be able to get very far. However, going downstream would be easy. Electromagnetic waves that have a short wavelength would be like if you had to paddle upstream for only a short distance. This is why higher frequency electromagnetic waves are able to pass through the earth's magnetosphere.

The sun's energy, in the form of electromagnetic waves that are able to pass through the earth's magnetosphere, affects the radio waves transmitted here on Earth. The amount of electromagnetic waves that strike the earth varies during the sun's eleven-year solar cycle. After sunset, the earth does not receive the sun-produced electromagnetic waves that affect the radio waves during the daytime. During the night, the low-frequency radio waves transmitted here on earth are able to travel along the earth's surface for great distances since their paths are no longer disturbed by the electromagnetic waves produced by the sun. This is why you are able to hear many distant radio stations at night.

During the day, when the electromagnetic waves produced by the sun are striking the earth, those waves also affect the paths of the radio waves transmitted here on Earth. This also is caused by the sun's electromagnetic waves which, like all electromagnetic waves, including radio waves, are composed of north and south magnetic monopoles. When these magnetic monopoles strike each other, the faster moving magnetic monopoles alter the paths of the slower moving magnetic monopoles.

During the day, you are not able to hear distant radio stations. This is a direct result of the faster moving magnetic monopoles in the form of electromagnetic waves produced by the sun.

The higher frequency radio waves transmitted here on Earth are unaffected, since at higher frequencies, the north and south magnetic monopoles that created the radio wave move much faster than the north and south magnetic monopoles that created the electromagnetic waves produced by the sun. However, during periods of high solar activity, when the sun releases more north and south magnetic monopoles as electromagnetic waves, the higher frequency radio waves transmitted here on earth are also affected.

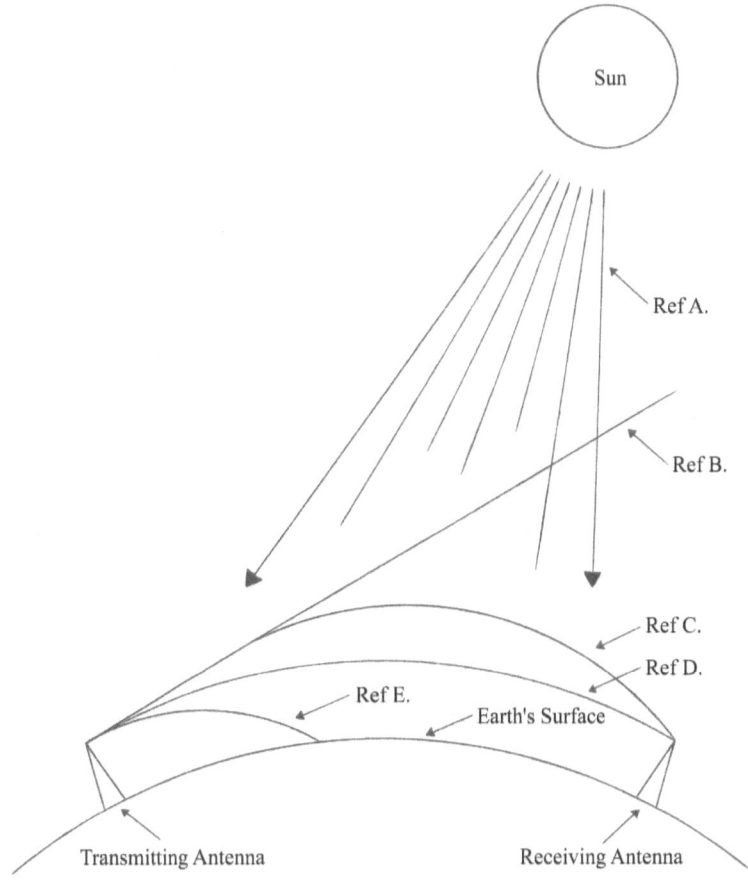

Fig 7 G.

Radio wave propagation

Figure 7 G, reference A shows the electromagnetic waves produced by the sun striking the earth's surface. These electromagnetic waves are created by the release of north and south magnetic monopoles during the nuclear reaction that is always taking place inside the sun. These electromagnetic waves are composed of the same north and south magnetic monopoles that radio waves are made of, and the faster moving monopoles will affect the way the radio waves travel here on earth.

Reference B shows that during the day, when solar activity is very low, the high-frequency radio waves are able to travel in a straight line because they are moving fast enough to be unaffected by the north and south magnetic monopoles coming from the sun and striking the earth. You can only hear these high-frequency radio waves for a short distance because these waves pass far above the receiving antenna.

Reference C shows that during the day, when solar activity is very high, the earth is being struck by more north and south magnetic monopoles in the form of electromagnetic waves than at night. This increase in magnetic monopoles is the result of the increase in sunspots. These monopoles are moving much faster than the magnetic monopoles that create the high-frequency radio waves, and this higher speed allows the monopoles to push the high-frequency radio waves closer to the earth's surface. This enables you to hear high-frequency radio waves that you would not normally be able to hear during periods of low solar activity. During low solar activity, these radio waves would have passed far above the receiving antenna, but during high solar activity, they are able to reach the receiving antenna due to the increase in the speed of the north and south magnetic monopoles striking the earth and altering the paths of the waves.

Reference D shows that after sunset, when the earth is no longer receiving the north and south magnetic monopoles that create the electromagnetic waves produced by the sun, low-frequency radio waves are able to travel along the earth's surface for great distances, enabling you to hear radio stations from far away.

Reference E shows that during the day, when the north and south magnetic monopoles are striking the earth, the path of the low-frequency radio waves is altered and pushed into the earth's surface, shortening their paths. This is the result of the faster moving magnetic monopoles coming from the sun and is why you are only able to hear the local radio stations during the daytime.

8. Superconductors

Superconductors are defined as any element, inter-metallic alloy, or compound that will conduct electricity without resistance and will repel a magnetic field. At the present, potential superconductor materials must be super-cooled to achieve the properties of a superconductor.

Before one can understand what makes a superconductor, one must thoroughly understand the atom, electricity, and magnetism, which have already been discussed in previous chapters. Recall that an atom has only three parts—north and south magnetic monopoles (which create the atom's magnetic lines of force and produces a magnetic dipole) and a particle of matter that represents one of the many elements from the periodic table.

The north and south magnetic monopoles attract each other and move in opposite directions, creating perpetual motion and thereby creating a magnetic line of force. This line of force is what produces the atom's magnetic field and thereby creates a magnetic dipole. The strength of this magnetic field changes with temperature.

When a bar magnet is heated to its Curie temperature, its magnetic field will begin to weaken. The same is true for the atoms' magnetic field when it's heated to a much higher temperature. It's this magnetic field that binds all atoms together, creating matter.

When the atoms' magnetic field becomes too weak, it is no longer able to bind one atom to another. This is why things turn into liquids when they are heated. On the other hand, when a bar magnet is cooled, its magnetic field becomes much stronger. Again, the same is true for the atoms' magnetic field. Under extremely low temperatures, all matter becomes solid.

Because the atom is only a small magnetic dipole, it acts the same as a compass needle, another magnetic dipole. The atom orients its south magnetic pole and points to the north magnetic pole of the bar magnet. When the atom is cooled to a low temperature, its magnetic field becomes stronger. This also creates stronger magnetic poles in the atom's magnetic field, which in turn produces a much stronger bond between the north and south magnetic poles of the atom. Because of this stronger bond between the atom's magnetic poles, they no longer respond to the magnetic lines of force from the bar magnet. Since the atom is only a small magnetic dipole, it no longer acts like a compass needle and follows the magnetic lines of force from an external magnetic field like a bar magnet; it instead points only to the magnetic poles of the atom that is adjacent to it.

Now imagine a row of compass needles that are all pointing to the earth's north pole. These compass needles are arranged so their needles all are perpendicular to each other. Now make the compass needles much stronger magnetic dipoles so they will point to their opposite magnetic poles and not the earth's magnetic pole. Because the compass needles all have a much stronger magnetic attraction for each other, they are no longer influenced by an outside magnetic force and will not point to the earth's north pole. The same thing happens to atoms in a superconductor when they are cooled to extremely low temperatures.

Remember from chapter 3 that electricity is generated by separating the north and south magnetic monopoles from atoms' magnetic lines of force into concentrated streams. These streams of magnetic monopoles also move in the same direction as the magnetic monopoles that created the atoms' magnetic lines of force. When these streams of north and south magnetic monopoles are recombined in a wire, they again produce a magnetic field since this is where the magnetic monopoles came from in the first place.

When the atoms in a superconductor are cooled to low temperature, their magnetic field becomes very strong. This strong magnetic field allows electricity to easily propagate through the atoms' magnetic lines of force and decreases its resistance to electricity. However, when the atoms' magnetic field becomes hot, it also becomes very weak, and electricity is not able to propagate as well through a weak magnetic field. The hot material has increased resistance to electricity.

Because electricity uses the atoms' magnetic lines of force to propagate, it does not propagate as well through a gas as through a metal. Since a gas has a much weaker magnetic field binding its atoms together, it also has a much higher resistance to electricity and thus requires a much higher voltage to overcome this resistance.

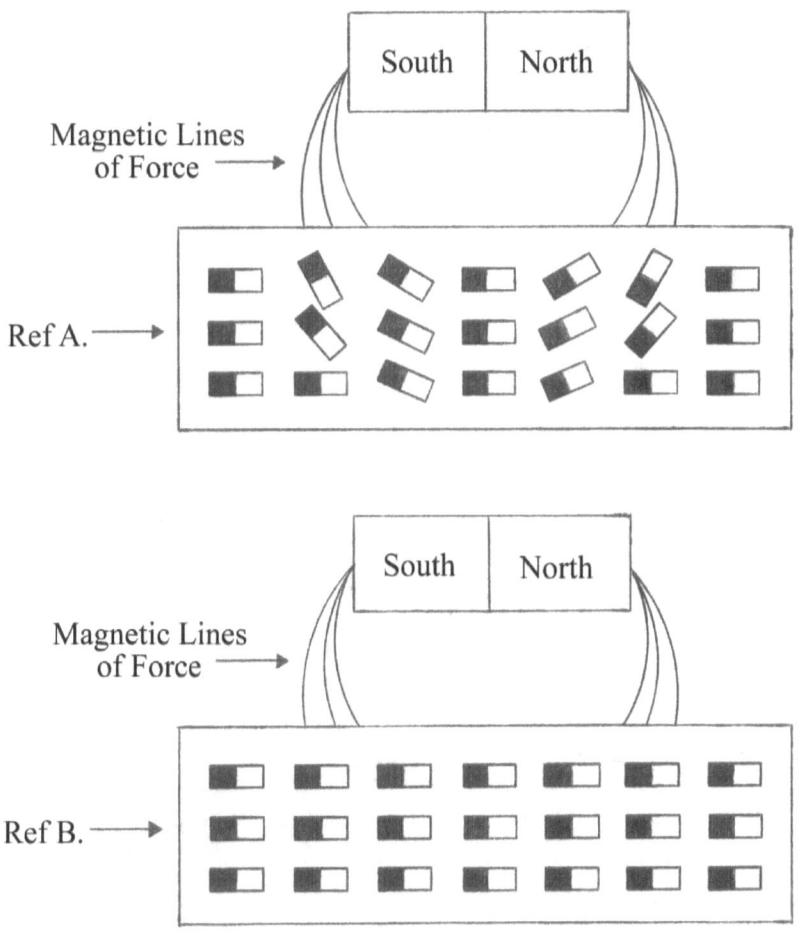

Fig 8.

A superconductor

Figure 8, reference A shows that the lines of force from an external magnetic field pass through the center of an atom's nucleus just like the earth's magnetic lines of force pass right through a compass needle and point to the earth's magnetic north pole. When a superconductor is cooled to extremely low temperatures, its atoms' magnetic field becomes very strong, binding the atoms together with a much stronger magnetic field and creating a stronger bond between the atoms' magnetic poles. In this state, the material will not be influenced by an external magnetic field.

Reference B shows that when superconductors are cooled to extremely low temperatures, their magnetic field becomes very strong. This strong magnetic attraction between atoms does not allow the lines of force from the external magnetic field to change their direction and pass through the atom's nucleus as shown in reference A. Since these magnetic lines of force are curved, they will intersect the atom's magnetic field at right angles and the atom's magnetic poles will repel the external magnetic lines of force. This is also known as the Meissner effect. This causes the external magnet to levitate above the superconductor because its magnetic lines of force are not able to pass through the superconductor. However, the external magnetic lines of force are able pass through the first row of atoms in the superconductor, an effect known as flux trapping. This allows the external magnetic field to attract the superconductor and repel it at the same time.

When the superconductor has again warmed to room temperature, the atoms' magnetic field weakens and again allows their magnetic poles to change their orientation and act like compass needles, following the magnetic lines of force from the external magnetic field. The external magnetic lines of force are again able to pass through the atoms' nuclei, and the superconductor no longer has its superconducting properties.

Superconductors also have very low resistance to electricity while they are super-cooled; this is direct result of the atoms' magnetic field. When the atoms in a superconductor are cooled to very low temperatures, the superconductor's magnetic field becomes very strong. Electricity uses the atoms' magnetic lines of force to propagate, and when an atom's magnetic field is very strong, it has little or no resistance to electricity. When the superconductor is again warmed to room temperature, its magnetic field once again becomes weaker and its resistance to electricity increases again. Remember, the stronger the atoms' magnetic lines of force are, the easier it is for electricity to propagate through. This is because the atoms' magnetic lines of force have north and south magnetic monopoles, which are moving in the same direction as the separated north and south magnetic monopoles that produce electricity. Remember, the cooler the temperature, the faster the perpetual motion of the atoms' magnetic lines of force. It's the speed of magnetic monopoles in the atoms' magnetic lines of force that offers less resistance to electricity.

Conclusion

All matter that makes up the universe contains only three particles—north and south magnetic monopoles and a particle of matter that represents an element. These are the building blocks of the universe. Magnetic monopoles create all forms of energy, from electromagnetic waves to magnetism and gravity. A simple experiment proves that magnetic monopoles exist, and that these monopoles can be separated and isolated into concentrated streams, creating electricity, simply using a bar magnet. Once these streams are recombined, they again create a magnetic field. When north and south monopoles move in the same direction relative to space, they create gravity. When they move opposite each other, taking the form of helices that can vary in diameter and frequency, electromagnetic waves are formed. When these monopoles move opposite each other in a continuous loop in perpetual motion, they create a magnetic dipole. These magnetic monopoles are what unite the fundamental forces of nature; they are the cosmic force that binds everything in the universe together.

Glossary

Alpha Particles: Particles that possess a positive/north charge and that are emitted from radioactive elements.

Aurora Borealis: A bright glow observed in the night sky, usually in the polar zone.

Beta Particles: Particles that possess a negative/south charge and that are emitted from radioactive elements.

Black Holes: Particles of matter that have collapsed into a large mass and that do not contain any north or south monopoles.

Centrifugal Force: The force in which a body revolving around a center tends to fly away from that center.

Covalent Bond: A process in which atoms bond using their magnetic lines of force.

Curie Temperature: The temperature above which a ferromagnetic material loses its permanent magnetism.

Dipole: A magnet that has both a north and a south pole.

Eddy Currents: Atoms that resist the reversal of their magnetic field, which is caused by an external magnetic field.

Electricity: The separation of magnetic monopoles into concentrated streams.

Electromagnet: A magnet that is created by passing electricity through a coil of wire.

Energy: The movement of magnetic monopoles.

Ether: Magnetic monopoles that are in north-south pairs and that are not moving relative to space.

Ferromagnetism: Phenomenon whereby materials retain their magnetic properties after an external magnetic field has been removed.

Flux Trapping: An effect of superconductors that does not allow an external magnetic field to penetrate more than a few atoms below the surface of the superconductor.

Gravity: Magnetic monopoles that are in north-south pairs and that are moving relative to space.

Heat: The random movement of magnetic monopoles, which causes the perpetual motion of atoms' magnetic lines of force, thereby slowing down and weakening their magnetic field.

Inertia: A tendency of all objects and matter in the universe to remain still or, if moving, to continue moving in the same direction unless acted on by another force.

Ionic Bond: A process in which atoms bond using additional monopoles that are attached to their magnetic lines of force.

Ions: Atoms with north or south magnetic monopoles attached to their opposite monopole in their magnetic lines of force.

Light: An electromagnetic wave that you are able to see. Composed of north and south monopoles.

Magnetism: The perpetual motion of north and south monopoles, which creates a dipole.

Matter: The bonding of two or more atoms together. Depending on the strengths of their magnetic field, the atoms create a solid, a liquid, or a gas.

Meissner Effect: The expulsion of a magnetic field from a superconductor.

Monopole: A particle that is a magnet with only one pole.

Néel Temperature: The temperature at which ferromagnetic and antiferromagnetic materials become paramagnetic.

Paramagnetism: A form of magnetism that occurs only in the presence of an externally applied magnetic field.

Quantization: The procedure of constraining something from a continuous set of values.

Superconductor: A substance that conducts an electric current with zero resistance.

About the Author

Lawrence Wippler has had a strong interest in astronomy and physics all his life. After high school, Wippler entered the military and served for thirteen years. During this time, he spent many hours in the library studying physics and astronomy, trying to get answers to the many unanswered questions he had about the universe. He decided he would begin a quest for a "Theory of Everything," a simple explanation that united the four fundamental forces of nature, which seemed to elude the physics community. His studies had led to a new way of thinking, of attacking the problem from a different perspective by tossing out the old rules of physics and starting fresh and clean. After leaving the military, he began working on his theory in earnest.

www.ingramcontent.com/pod-product-compliance
Lightning Source LLC
Chambersburg PA
CBHW030852180526
45163CB00004B/1546